Applying
Generalizability
Theory
using EduG

QUANTITATIVE METHODOLOGY SERIES

George A. Marcoulides, Series Editor

This series presents methodological techniques to investigators and students. The goal is to provide an understanding and working knowledge of each method with a minimum of mathematical derivations. Each volume focuses on a specific method (e.g. Factor Analysis, Multilevel Analysis, Structural Equation Modeling).

Proposals are invited from interested authors. Each proposal should consist of: a brief descriptions of the volume's focus and intended market; a table of contents with an outline of each chapter; and a curriculum vita. Materials may be sent to Dr. George A. Marcoulides, University of California—Riverside, george.marcoulides@ucr.edu.

Applying Generalizability Theory using EduG

Jean Cardinet
IRDP, Neuchâtel
Switzerland

Sandra Johnson
Assessment Europe
Scotland

Gianreto Pini
University of Geneva
Switzerland

Routledge
Taylor & Francis Group
New York   London

Routledge
Taylor & Francis Group
711 Third Avenue
New York, NY 10017

Routledge
Taylor & Francis Group
27 Church Road
Hove, East Sussex BN3 2FA

International Standard Book Number: 978-1-84872-828-8 (Hardback) 978-1-84872-829-5 (Paperback)

Library of Congress Cataloging-in-Publication Data

Cardinet, Jean.
  Applying generalizability theory using EduG / Jean Cardinet, Sandra
Johnson, Gianreto Pini.
      p. cm. -- (Quantitative methodology series)
  Includes bibliographical references and index.
  ISBN 978-1-84872-828-8 (hardcover : alk. paper) -- ISBN 978-1-84872-829-5
(pbk. : alk. paper)
    1. Educational statistics--Evaluation. 2. Educational tests and
measurements--Validity. 3. Curriculum evaluation. 4. Analysis of variance. I.
Johnson, Sandra, 1946- II. Pini, Gianreto. III. Title.

LB2846.C28 2010
370.7'27--dc22
                                                                    2009034379

Visit the Taylor & Francis Web site at
http://www.taylorandfrancis.com

and the Psychology Press Web site at
http://www.psypress.com

Contents

SOFTWARE NOTICE

An algorithmic error has been detected in *EduG* version 5, which was used to produce the analysis results presented and discussed in this book. Components of variance for interactions involving both fixed and random facets are not correctly computed. This does not in any way detract from the value of the book as a tutorial on the principles and practice of generalizability theory.

The software problem has been resolved in version 6, which is available for download under the name *EduG 6.0 e*, from the following web address:

http://www.irdp.ch/edumetrie/englishprogram.htm

You should download and use version 6 for any future work that you do with G theory. But continue to use *EduG* version 5, which is now available as *PREVIOUS EduG* from the same web address, whenever you work through the practice examples in Chapter 5.

Foreword

Applying Generalizability Theory Using EduG concisely and understandably conveys a flexible approach to comprehending and modeling behavioral measurements. It breaks with tradition. And as will be described below, it conveys important, if somewhat contested, advances in "generalizability theory." The book does so in its pages, and in its accompanying user-friendly software package, EduG. To see why all this is so, read on!

Generalizability (G) theory is a "grand theory" of the dependability or reliability of behavioral measurements. It models, conceptually, the way in which such measurements are made. This is accomplished by identifying as many potential sources contributing to variation in scores as possible. And it provides statistical estimates of the magnitude of these sources of variation in scores. For example, behavioral measurement often focuses on individuals' performance across a set of test items or job tasks as scored by two or more judges. So, conceptually, individuals, items/tasks, and judges can contribute variation among a set of scores. Some of this variation is expected and wanted, such as systematic variation among students' scores on a science achievement test or systematic variation among incumbents' scores on a job-performance test. Of course, some of the variation is unwanted, such as inconsistency in students' performance from one science test item to another or inconsistency in job incumbents' performance from one job task to another. G theory provides estimates of the magnitude of wanted variation—true-score or universe-score variation—and unwanted variation—error variation. It does so for each source of variation—individuals, items/tasks, and judges—and all their possible combinations.

G theory provides these estimates within a statistical sampling framework by assuming that a behavioral measurement is a *sample* of behavior drawn from a large universe of behavior. Ideally we would like to know an individual's average score, for example, on all possible tasks as evaluated by all possible judges. However, this is impractical, if not impossible.

Rather, in practice we sample a few items/tasks and judges and calculate an average sample score for that individual. We then generalize from this sample score to what this individual's average score would be in the larger universe of interest. G theory, then, permits us to address such questions as: Is the sampling of tasks or judges the major source of measurement error? Can we improve the reliability of the measurement further by increasing the number of tasks or the number of judges, or is some combination of the two more effective? Are the test scores adequately reliable to make decisions about the level of a person's performance for a certification decision or for rank ordering individuals in order to select the top scorers for a limited number of positions available on a job or the number of openings in a college?

G theory estimates the sources of variation among scores within the analysis of variance framework. The analysis of variance divides up the total variation among scores statistically into its component parts. So, in the example above, the analysis of variance would partition scores into the effects of individuals, items/tasks, and judges—and their two-way combinations (individuals × items/tasks, individuals × judges, items/tasks × judges) and their three-way combination (individuals × items/tasks × judges). For each source of variation, the theory and analysis of variance provide an estimate of its magnitude. Large variation among individuals—true- or universe-score variance in the example—is wanted and expected; it contributes to the reliability of the measurement. Large variation in the other sources of variation and all the combinations is unwanted as these sources of variation contribute uncertainty or error in the measurement, thereby decreasing reliability.

These estimates of variation enable us to tailor a measurement in such a way as to maximize its reliability under cost and time constraints. If variation among items on average and variation in particular in individuals' performance on different items introduce large error variation into scores, then this unwanted variation can be countered by increasing the sample size—increasing the number of items/tasks in the measurement procedure. In this way, the test user can ask "what if" questions such as those raised above: What if I increase the number of tasks and minimize the number of raters—will that give me the measurement reliability I need at low cost?

At this point, if you haven't decided to move immediately to the book's Preface, you might be wondering why I'm telling you all this! Well, in order to understand the contribution *Applying Generalizability Theory Using EduG* makes, and how it fits into the larger G theory picture, some background is needed.

Applying Generalizability Theory Using EduG is a wonderful book for many reasons; I enumerate three. It provides in one place the fruits of

25 years of research carried out by its authors, especially Jean Cardinet (as well as others, of course, most notably Lee Cronbach and his collaborators). Cardinet's contributions are quite important and special (and some a bit controversial!). It is important that the world has ready access to his and his co-authors' ideas. Second, the book is an "easy read," at least compared to other books on G theory such as Cronbach et al.'s *Dependability of Behavioral Measurements* and Bob Brennan's *Generalizability Theory*. I say this because the conceptual and statistical ideas are presented clearly; they are immediately exemplified in concrete terms. Moreover, I say this because I have used a draft of the book in my course on G theory at Stanford and graduate students have universally found the available chapters readable and clear. Third, the book comes with outstanding downloadable software—EduG—for carrying out G theory analyses. It explains the use of this software thoroughly and understandably, again with extensive, concrete examples. The software is easy to use, contains important options for analyzing measurement data not available elsewhere, and provides output that is very readable. To be sure, the software does not do everything (e.g., it does not implement multivariate G theory) but it does a good 90% of everything and close to 100% of what is ordinarily needed. Importantly, it does some things no other G theory software package does, as we will see below.

There are certain notable aspects of G theory presented in this book that are not readily available in other books on G theory (Cronbach et al.'s, Brennan's, or the *Primer* Noreen Webb and I wrote). One of the most important contributions that Cardinet and his colleague Tourneur made was to recognize that the same behavioral measurement might be used for more than one purpose. For example, the user of a job performance test might want to evaluate the performance of individuals. So, systematic differences among individuals on the test represent wanted variance in scores—individuals vary in their performance on the job. Variation among individuals' performance from one task to another, or from one judge to another, "muddies the water"—that is, it obscures systematic individual differences in performance and thus is unwanted variation. In contrast, another, test user might be interested in the variation in the difficulty of job tasks. In this case, variation in the level of job task scores (averaged over individuals and judges) would be wanted variation. And variation among students' performances across tasks and judges would muddy the water—that is, contribute error to the measurement.

Cardinet and Tourneur called this flexibility the *symmetry* of a behavioral measurement. The analysis of variance is indifferent to what the test user calls "the object of measurement" and "measurement error." It estimates the magnitude of variation attributable to each possible source of variation (that the measurement design permits). It is, then, up to the user to decide what is wanted variation and what is error variation. This seemingly simple but extraordinarily powerful insight permitted Cardinet and Tourneur to presage developments in G theory by others not necessarily aware of their symmetry notion. So, for example, estimation of the reliability of (mean) scores when classrooms, not individual students, are the object of achievement measurement needed no special treatment. Variation in classrooms was wanted variation and variation among individuals and judges was unwanted variation. Symmetry is implemented directly in EduG; this is one of the most important contributions of the software.

A closely related notion to symmetry is the distinction Cardinet and Tourneur made between what they called "the face of differentiation" and "the face of instrumentation." The former refers to the object of measurement and the latter refers to those sources of variation in the measurement that contribute to error. The two "faces" change with the change in the use of the measurement—in our example, one user focused on variation among individuals while another focused on job-task variation. While Cronbach and others recognized that the face of instrumentation might have a complex structure (e.g., test items nested within content domains), Cardinet and Tourneur recognized that this could also be true of the face of differentiation. For example, we might have, as the object of measurement, students nested within their respective genders: male and female.

Once we admit to complex objects of measurement as well as complex instrumentation, we've let the cat out of the bag, the fox into the chicken coop, the camel's nose under the tent, well, you get what I mean. To understand why this becomes an important issue, bear with me, as I need to make another historical distinction.

G theory grew out of classical test theory. This theory focused on individual differences, especially among people, and on the reliability of a measurement for rank ordering those individuals. In its most general form, the simple assumption was that individuals were randomly sampled from a population of people and that (say) tasks were randomly sampled from a universe of possible tasks. Building on this earlier work, G theory is, essentially, a "random effects" theory where observations are selected randomly from an indefinitely large universe. And that's the way Cronbach and his collaborators saw it. However, male and female are not random samples from an indefinitely large "gender population." Nor are content areas on an achievement test (e.g., physics, chemistry, biology, earth science). Rather, gender exhausts the two levels of the population and the four content areas exhaust the science content universe. Gender and content domain, then, are fixed in the sense that all levels of the population and of the universe are observed, respectively, not sampled. Cronbach and colleagues' solution to this "problem" was to avoid nested faces of differentiation: "We shall give no more than incidental attention to designs in which subjects are nested" (Cronbach, Gleser, Nanda, & Rajaratnam, 1972, p. 35). They treat a fixed facet in the Face of Instrumentation in one of three ways: (1) average over the levels of a fixed source of variation, (2) examine variation among other sources of variation at each level of the fixed source (variation for males and variation for females separately), or (3) view the levels of a fixed facet as a set of "multivariate outcomes." Hence, G theory remained "essentially" a random-effects theory.

This is where Cardinet and Tourneur parted company from mainstream (American) G theory. They reasoned that variation from random and fixed sources was both meaningful to a measurement and should be incorporated into G theory. That is, they argued that G theory should be "essentially" a mixed-effects (random and fixed) theory. Consequently, if the face of differentiation included students nested within male and female, then student to student variation should be estimated as part of the variation in the object of measurement *as should the mean differences in scores between males and females*. Once fixed effects are admitted to G theory, the theory expands greatly, but not without controversy. As I said before, I fall to a large degree in the Cardinet camp on this, even though Cronbach was my dissertation advisor and close friend for nearly 35 years.

Cardinet parted company with the mainstream as well with certain aspects of popular statistical estimation theory, that is, on how to estimate the sources of variation in the measurement. This may sound exotic on the one hand and dangerous on the other hand. But it isn't really. Do you recall that when you were learning statistics, you were told that the standard deviation was calculated by dividing by $N - 1$ and not N? You probably asked why and got an unintelligible answer. Well, Cardinet and Tourneur would have said yes, you can divide by $N - 1$, that's one way of doing it. But you can also use N and it is a matter of definition. The book's authors, then, chose to "divide by N" and in so doing made many of the computations in their "mixed-effects" G theory readily tractable. In Cardinet's words, "Our choice has the advantage of giving a model with which we can combine easily facets sampled in the three possible ways (randomly from an infinite universe, randomly from a finite universe, and fixed with a set number of levels included), on both the differentiation and the instrumentation face. The other choice [$N - 1$], as far as we know, is not practical to treat these problems in G theory" (Cardinet, email March 6, 2008).

In the end, the use of N and not $N - 1$ in most cases doesn't matter at all. For example, when the face of differentiation contains a single random object-of-measurement facet and a simple or highly complex face of instrumentation, Cardinet's theory and Cronbach's (and Brennan's, and Shavelson and Webb's) equations and results from software (EduG and Brennan's GENOVA) are identical. In "nonstandard" designs where the face of differentiation is complex with a fixed facet, only Cardinet's theory treats this, producing a kind of omega squared coefficient (a kind of effect size). However, the authors recognized the potential danger in using biased variance component estimates (dividing by N) that arises from their general approach to estimating fixed and random effects similarly. They appropriately warn the reader when it is essential to use unbiased estimates (dividing by $N - 1$). The flexibility and power gained by the approach espoused in *Applying Generalizability Theory Using EduG*, then, seem to outweigh the *au courant* view.

Once the book's authors had made this turn away from popular tradition, and justifiably in my reckoning, using a mixed-effects model for G theory became possible and the variation in both the face of differentiation and in the face of instrumentation due to random *and* fixed sources became possible. EduG implements this insight and provides a comprehensive treatment of behavioral measurement including measurements with random and fixed sources of variation.

Well, this Foreword got a bit longer than intended. But I felt it important to point out some of the major contributions to G theory that are embodied in *Applying Generalizability Theory Using EduG*. I hope you will enjoy and use the book as much as I have and intend to in the future.

Richard J. Shavelson
Margaret Jacks Professor of Education,
School of Education, Stanford University

Preface

This book, like G theory itself, is the product of a long history of international cooperation. We, the three authors, are ourselves of different nationalities. We live too far apart to be able to meet regularly face to face, and so, as is commonplace these days, we have used the Internet to communicate with each other as we have produced this text. What brings us together is the fact that each one of us has had a long personal involvement with G theory, which in one case reaches back to the very genesis of the methodology itself.

In the early 1960s, as a research associate at the University of Illinois, Urbana, Jean Cardinet worked with Lee Cronbach as he and his colleagues were in the throes of developing G theory. It was this earliest introduction to the generalizability concept that inspired his lifelong interest in this particular methodology. But it was the publications of Robert Brennan (especially his graciously forwarded prepublication papers) that equipped Jean Cardinet with the statistical background necessary to extend the theory further. In 1985 he published, with Yvan Tourneur of the University of Mons in Belgium, a text that has become a seminal work on G theory in the French-speaking world—*Assurer la Mesure* (Berne: Peter Lang). His theoretical interest was sustained over the following years by stimulating intellectual exchanges with colleagues in Europe (Linda Allal among them) and the United States (including Richard Shavelson and George Marcoulides).

For her part, Sandra Johnson was one of the earliest G theory practitioners. During the two years following the publication of Cronbach's seminal 1972 text—*The Dependability of Behavioral Measurements* (New York: Wiley)—she worked as a research associate at the University of Manchester, England, exploring the reliability of proposed new forms of school examinations. She extended her experience of G theory as she collaborated with research colleagues in establishing a sample-based national attainment survey program in science in the 1980s, based at the University of Leeds. She further developed the model when evaluating a technique for comparing examination standards, which involved a fixed facet object of measurement. It was at this point, in the mid-1980s, that she and Jean Cardinet began their long professional association.

Gianreto Pini's history of involvement with G theory began in the late 1980s. On the untimely death of Yvan Tourneur, Jean Cardinet's Belgo-Swiss collaboration ended. But a newly established working group (Groupe Edumetrie) of the Swiss Society for Research in Education (SSRE) took up the cudgel, under the direction of Daniel Bain. Gianreto Pini has been a key methodological contributor to this group from its inception. Assisted by his experience of teaching G theory at the University of Geneva, in 1996 he and Daniel Bain coauthored the G theory primer *Pour évaluer vos évaluations—La généralisabilité: mode d'emploi* (Geneva Centre for Psycho-Educational Research). It is through his long involvement in the Groupe Edumetrie that his personal association with Jean Cardinet has not only endured but strengthened over time.

The lack of availability of user-friendly flexible software has for many years proved an obstacle to the widespread application of G theory. In the early 1980s, in an attempt to address this situation, François Duquesne, a Belgian colleague, developed a program called ETUDGEN, written in Basic to run on an Apple II (48k). This to some extent served the essential needs of academic practitioners, but it quickly proved too limited for several kinds of applications. His work was taken up and further developed by colleagues at the Center for the Evaluation of Health Sciences in the University of Laval (CESSUL), who, in the 1990s, put ETUDGEN on the Internet, in both English and French.

But practitioners needed PC-based software. In response, Pierre Ysewijn in Switzerland started work on "GEN," but was unable to complete the project due to lack of financial support. His work was taken up eventually by the *Groupe Edumetrie*, which managed to raise the necessary financial resources to be able to invite Educan, a Montreal-based software company, to take over program development. And so EduG, the first genuinely versatile and user-friendly G theory software package, was born. Jean Cardinet oversaw the software design, aided by members of *Groupe Edumetrie*, and in particular by Daniel Bain and Gianreto Pini. Jean Cardinet and Sandra Johnson together collaborated on producing the Help pages and the User Guide in French and in English. In 2006, EduG was finally made available to users internationally, as freeware downloadable from the website of the Institute for Educational Research and Documentation (IRDP) in Neuchatel, Switzerland: http://www.irdp.ch/edumetrie/englishprogram.htm.

EduG began to be used by colleagues in Europe and the United States, and so it was George Marcoulides who invited Jean Cardinet to produce this book for his Routledge/Taylor & Francis Collection (Quantitative

Methodology Series), a book that would not only illustrate applications of the software for interested readers and potential practitioners, but which would also describe its theoretical foundation. Jean Cardinet in turn invited Gianreto Pini and Sandra Johnson to join him in this venture. The collaboration has been challenging, frustrating at times, but above all enjoyable. We hope that what we have produced will be of interest and value to many readers around the world.

But exactly what was our aim for this book? First and foremost it was to show how G theory can usefully be applied in many different contexts, not only in education and psychology, but also elsewhere. Thus far, G theory has principally been applied in the field of student testing, with the result that it has been in danger of being seen as merely a refinement of the psychometrics discipline. We had other ambitions for the methodology. We thought that it has the potential to explore and to assure the quality of *any* kind of measurement, in whatever experimental discipline, always comparing the size of "true" variance with that of variance attributable to "noise," that is, to nonvalid systematic and random influences on measurements. The calculation of effect size in experimental design applications is a particularly important facility offered by G theory.

We have tried our best to produce a pedagogically effective text. And so, in order to maximize understanding on the part of readers, fundamental principles are explained on the basis of concrete examples. We have also included a section at the end of the book explaining key terms; these are referenced from within the text through use of superscript identifiers. Since learning is reinforced through independent practice, it is in the same spirit that we have ensured that the data sets associated with the commented example applications described in Chapter 4 are supplied with EduG, so that interested readers will be able to repeat the analyses themselves. Additional application examples also have their associated data sets downloadable from http://www.psypress.com/applied-generalizability-theory or the authors' websites. We leave it to the new EduG user to prepare the analyses and interpret the results, before consulting the guidance support that is also included in the book.

Numerous teams have contributed to the development of G theory, at first on both sides of the Atlantic, then in various languages, and within different cultures, but now on a completely international level, as extended in scope as the Internet itself. We are far from the view that we have ourselves identified, least of all resolved, all the issues concerning the theory that we enthusiastically promote. And we are ready to acknowledge that other approaches, in particular that of item response theory, are capable of playing complementary roles, which we would equally like to see further developed. We particularly look forward to a future unification of the different approaches within a single theory of measurement.

Theoretical and applied research in this field will continue, prolonging in space and time this inspiring chain of cooperation between countries and between generations. We hope that this book will prove an active stimulant in this endeavor.

Jean Cardinet
Sandra Johnson
Gianreto Pini

Acknowledgments

First and foremost, we would like to acknowledge Lee Cronbach, the originator of G theory, and Robert Brennan, who further developed the theory, particularly in terms of mixed model ANOVAs. Both have inspired our work in this particular field.

Next we thank George Marcoulides and Richard Shavelson, for so effectively encouraging us to produce this book, providing a valuable opportunity to document several developments that had remained unpublished in English up to now.

Numerous other colleagues, in Switzerland, Belgium, the United Kingdom, Canada, and the United States, also deserve our thanks for having stimulated our efforts with their continuing interest. Daniel Bain is foremost among them. We are grateful to him firstly for having ensured the development of EduG, which was a major objective of the special interest group that he headed (Édumétrie—Qualité de l'évaluation en éducation, of the Swiss Society for Educational Research). We are also indebted to him for agreeing to use, in Chapter 4, several application examples that originally appeared in a French-language publication produced by him in collaboration with Gianreto Pini, along with the use, in Chapter 5, of examples of his own work (some of his detailed research reports are available in French at the following address: http://www.irdp.ch/edumetrie/exemples.htm). We express our sincere gratitude to Daniel Bain for his important contribution to the preparation of this book, which would not have been realized without the multifaceted help of his long-standing collaboration.

We would like to offer special acknowledgments to Linda Allal, who contributed to the very first English-language publications on the symmetry of G theory and who is also the one personal link connecting we three authors. Each of us has participated over the years in one or other of her numerous activities at the University of Geneva. Directly and indirectly, this book owes a lot to her. Maurice Dalois and Léo Laroche of

EDUCAN deserve our thanks for their collaboration over many years in the development of EduG, and in particular for their unwavering willingness to introduce modifications even as this book was being drafted. In regard to the book itself, we would like to thank the reviewers Dany Laveault from the University of Ottawa and Robert Calfee from Stanford University for their very helpful and much appreciated comments.

Finally, we offer huge thanks to our partners, families, and friends. What a lot of patience they had, when for countless days over a long period of time family and social lives were sacrificed in the interests of our hours of solitary writing. We fervently hope that the end has to some extent justified the means.

chapter one

What is generalizability theory?

Generalizability theory: Origin and developments

Not all measuring procedures can be perfectly accurate. In the social and health sciences in particular, but in the natural sciences as well, we can rarely assume our measurements to be absolutely precise. Whether we are attempting to evaluate attitudes to mathematics, managerial aptitude, perception of pain, or blood pressure, our scores and ratings will be subject to measurement error. This is because the traits or conditions that we are trying to estimate are often difficult to define in any absolute sense, and usually cannot be directly observed. So we create instruments that we assume will elicit evidence of the traits or conditions in question. But numerous influences impact on this process of measurement and produce variability that ultimately introduces errors in the results. We need to study this phenomenon if we are to be in a position to quantify and control it, and in this way to assure maximum measurement precision.

Generalizability theory, or G theory, is essentially an approach to the estimation of measurement precision in situations where measurements are subject to multiple sources of error. It is an approach that not only provides a means of estimating the dependability of measurements already made, but that also enables information about error contributions to be used to improve measurement procedures in future applications. Lee Cronbach is at the origin of G theory, with seminal co-authored texts that remain to this day essential references for researchers wishing to study and use the methodology (Cronbach, Gleser, Nanda, & Rajaratnam, 1972; Cronbach, Rajaratnam, & Gleser, 1963).

The originality of the G theory approach lies in the fact that it introduced a radical change in perspective in measurement theory and practice. In essence, the classical correlational paradigm gave way to a new conceptual framework, deriving from the analysis of variance (ANOVA), whose fundamental aim is to partition the total variance in a data set into a number of potentially explanatory sources. Despite this profound change in perspective, G theory does not in any way contradict the results and contributions of classical test theory. It rather embraces them as special cases in a more general problematic, regrouping within a unified conceptual framework concepts and techniques that classical theory presented in a disparate, almost disconnected, way (stability, equivalence, internal consistency, validity, inter-rater agreement, etc.). The impact of the change in perspective is more than a straightforward theoretical reformulation. The fact that several identifiable sources of measurement error (markers, items, gender, etc.) can simultaneously be incorporated into the measurement model and separately quantified means that alternative sampling plans can be explored with a view to controlling the effects of these variables in future applications. G theory thus plays a unique and indispensable role in the evaluation and design of measurement procedures.

That is why the *Standards for educational and psychological testing* (AERA, 1999), developed jointly by the American Educational Research Association, the American Psychological Association, and the National Council on Measurement in Education (and hence familiarly known as the "Joint Standards"), stress the need to refer to G theory when establishing the validity and reliability of observation or testing procedures. The first two chapters immediately embrace this inferential perspective, in which generalization to a well-defined population is made on the basis of a representative random sample. The Standards explicitly refer to G theory at several points. For instance, the commentary for standard 2.10 states, with respect to reliability estimates based on repeated or parallel measures:

> Where feasible, the error variances arising from each source should be estimated. Generalizability studies and variance component analyses are especially helpful in this regard. These analyses can provide separate error variance estimates for tasks within examinees, for judges and for occasions within the time period of trait stability. (AERA, 1999, p. 34)

We return later to some of the essential characteristics of the theory. For the moment we simply draw attention to two important stages in its evolution, of which the second can be considered as an extension of the first, since it has led to the expansion and considerable diversification of its fields of application.

As originally conceived, G theory was implicitly located within the familiar framework of classical test theory, a framework in which individuals (students, psychiatric patients, etc.) are considered as the objects of measurement, and the aim is to differentiate among them as reliably as possible. The principal requirement is to check that the instrument to be used, the test or questionnaire, can produce reliable measurements of the relative standing of the individuals on some given measurement scale, despite the inevitably disturbing influence on the measures of the random selection of the elements of the measurement instrument itself (the test or questionnaire items).

During the 1970s and 1980s, a potentially broader application of the model was identified by Jean Cardinet, Yvan Tourneur, and Linda Allal, who observed that the inherent symmetry in the ANOVA model that underpinned G theory was not being fully exploited at that time. They noted that in Cronbach's development of G theory the factor Persons was treated differently from all other factors, in that persons, typically students, were consistently the only objects of measurement. Recognizing and exploiting model symmetry (i.e., the fact that any factor in a factorial design has the potential to become an object of measurement) allows research procedures as well as individual measurement instruments to be evaluated. Thus, procedures for comparing subgroups (as in comparative effectiveness studies of various kinds) can also be evaluated for technical quality, and improved if necessary (Cardinet & Allal, 1983; Cardinet & Tourneur, 1985; Cardinet, Tourneur, & Allal, 1976, 1981, 1982). As these authors were expounding the principle of model symmetry, practitioners on both sides of the Atlantic were independently putting it into operation (e.g., Cohen & Johnson, 1982; Gillmore, Kane, & Naccarato, 1978; Johnson & Bell, 1985; Kane & Brennan, 1977).

Relative item difficulty, the mastery levels characterizing different degrees of competence, the measurement error associated with estimates of population attainment, the progress recorded between one stage and another within an educational program, the relative effectiveness of teaching methods, are all examples of G theory applications that focus on something other than the differentiation of individuals. To facilitate an extension to the theory, calculation algorithms had to be modified or even newly developed. Jean Cardinet and Yvan Tourneur (1985), whose book on G theory remains an essential reference in the French-speaking world, undertook this task. We explicitly place ourselves in the perspective adopted by these researchers.

An example to illustrate the methodology

The example

It will be useful at this point to introduce an example application to illustrate how G theory extends classical test theory, and in particular how the principle of symmetry enriches its scope. Let us suppose that a research study is planned to compare the levels of subject interest among students

taught mathematics by one or the other of two different teaching methods, Method A and Method B. Five classes have been following Method A
and five others Method B. A 10-item questionnaire is used to gather the
research data. This presents students with statements of the following
type about mathematics learning:

- Mathematics is a dry and boring subject
- During mathematics lessons I like doing the exercises given to us in
  class

and invites them to express their degree of agreement with each statement, using a 4-point Likert scale. Students' responses are coded numeri-
cally from 1 (*strongly agree*) to 4 (*strongly disagree*), and where necessary
score scales are transposed, so that in every case low scores indicate low
levels of mathematics interest and high scores indicate high levels of
mathematics interest. There are then two possibilities for summarizing
students' responses to the 10-item questionnaire: we can sum students'
scores across the 10 items to produce total scores on a 10–40 scale (10 items,
each with a score between 1 and 4), or we can average students' scores
over the 10 items to produce average scores on a 1–4 scale, the original
scale used for each individual item. If we adopt the second of these alter-
natives, then student scores higher than 2.5, the middle of the scale, indicate positive levels of interest, while scores below 2.5 indicate negative
levels of interest; the closer the score is to 4, the higher the student's general mathematics interest level, and the closer the score is to 1 the lower
the student's general mathematics interest level.

Which reliability for what type of measurement?

As we have already mentioned, the aim of the research study is to compare two mathematics teaching methods, in terms of students' subject
interest. But before we attempt the comparison we would probably be
interested in exploring how "fit for purpose" the questionnaire was in
providing measures of the mathematics interest of individual students.
Of all the numerous indicators of score reliability developed by different individuals prior to 1960, Cronbach's α coefficient[i] (Cronbach, 1951)
remains the best known and most used (Hogan, Benjamin, & Brezinski,
2000). The α coefficient was conceived to indicate the ability of a test to
differentiate among individuals on the basis of their responses to a set of
test items, or of their behavior within a set of situations. It tells us the
extent to which an individual's position within a score distribution
remains stable across items. α coefficients take values between 0 and 1; the
higher the value, the more reliable the scores. The α value in this case is

0.84. Since α values of at least 0.80 are conventionally considered to be acceptable, we could conclude that the questionnaire was of sufficient technical quality for placing students relative to one another on the scale of measurement. This is correct. But in terms of what we are trying to do here—to obtain a reliable measure of average mathematics interest levels for each of the two teaching methods—does a measure of internal consistency, which is what the α coefficient is, really give us the information we need about score reliability (or score precision)?

We refer to the Joint Standards again:

> ... when an instrument is used to make group judgments, reliability data must bear directly on the interpretations specific to groups. Standard errors appropriate to individual scores are not appropriate measures of the precision of group averages. A more appropriate statistic is the standard error of the observed score means. Generalizability theory can provide more refined indices when the sources of measurement are numerous and complex. (AERA, 1999, p. 30)

In fact, the precision, or rather the imprecision, of the measure used in this example depends in great part on the degree of heterogeneity among the students following each teaching method: the more heterogeneity there is the greater is the contribution of the "students effect" to measurement error. This is in contrast with the classical test theory situation where the greater the variance among students the higher is the "true score" variance and consequently the higher is the α value. Within-method student variability is a source of measurement error that should not be ignored. Moreover, other factors should equally be taken into consideration: in particular, variability among the classes (within methods), variability among the items, in terms of their overall mean scores, as well as any interactions that might exist between teaching methods and items, between students (within classes) and items, and between classes and items.

How does G theory help us?

As we will show, G theory is exactly the right approach to use for this type of application. It is sufficient to consider the two teaching methods as the objects of measurement and the other elements that enter into the study (items, students, and classes) as components in the measurement procedure, "conditions of measurement," potentially contributing to measurement error. In place of the α coefficient we calculate an alternative reliability indicator, a generalizability coefficient (G coefficient). Like the α coefficient, G coefficients are variance ratios. They indicate the proportion of total score variance that can be attributed to "true" (or "universe") score variance, which in this case is inter-method variation, and equivalently the proportion of variance that is attributable to measurement error. Also like α, G coefficients take values between 0 (completely unreliable measurement) and 1 (perfectly reliable measurement), with 0.80 conventionally accepted as a minimum value for scores to be considered acceptably reliable. The essential difference between measurement error as conceived in the α coefficient and measurement error as conceived in a more complex G coefficient is that in the former case measurement error is attributable to one single source of variance, the student-by-item interaction (inconsistent performances of individual students over the items in the test), whereas in the latter case multiple sources of error variance are acknowledged and accommodated.

A G coefficient of *relative* measurement indicates how well a measurement procedure has differentiated among objects of study, in effect how well the procedure has ranked objects on a measuring scale, where the objects concerned might be students, patients, teaching methods, training programs, or whatever. This is also what the α coefficient does, but in a narrower sense. A G coefficient of *absolute* measurement indicates how well a measurement procedure has located objects of study on a scale, irrespective of where fellow objects are placed. Typically, "absolute" coefficients have lower values than "relative" coefficients, because in absolute measurement there are more potential sources of error variance at play. In this example, with 15 students representing each of the five classes, the relative and absolute G coefficients are 0.78 and 0.70, respectively (see Chapter 3 for details). This indicates that, despite the high α value for individual student measurement, the comparative study was not capable of providing an acceptably precise measure of the difference in effectiveness of the two teaching methods in terms of students' mathematics interest.

In this type of situation, a plausible explanation for low reliability can sometimes be that the observed difference between the measured means is particularly small. This is not the case here, though: the means for the two teaching methods (A and B) were, respectively, 2.74 and 2.38 (a difference of 0.36) on the 1–4 scale. The inadequate values of the G coefficients result, rather, from the extremely influential effect of measurement error, attributable to the random selection of small numbers of attitude items, students, and classes, along with a relatively high interaction effect between teaching methods and items (again, Chapter 3 provides the details).

Standard errors of measurement,[ii] for relative and for absolute measurement, can be calculated and used in the usual way to produce confidence intervals[iii] (but note that adjustments are sometimes necessary, as explained in Chapters 2 and 3). In this example, the adjusted standard errors are equal to 0.10 and 0.11, respectively, when the mean results of the two teaching methods are compared. Thus a band of approximately two standard errors (more specifically 1.96 standard errors, under Normal distribution assumptions) around each mean would have a width of approximately ±0.20 for relative measurement and ±0.22 for absolute measurement. As a result, the confidence intervals around the method means would overlap, confirming that the measurement errors tend to blur the true method effects.

Optimizing measurement precision

Arguably the most important contribution of the G theory methodology, and the most useful for practitioners wanting to understand how well their measuring procedures work, is the way that it quantifies the relative contributions of different factors and their interactions to the error affecting measurement precision. G coefficients are calculated using exactly this information. But the same information can also be used to explore ways of improving measurement precision in a future application. In the example presented here, the principal sources of measurement error were found to be inter-item variation, inter-class (within method) variation, and inter-student (within class) variation. The interaction effect between methods and items also played an important role. Clearly, the quality of measurement would be improved if these major contributions to measurement error could be reduced in some way. A very general, but often efficient, strategy to achieve this is to use larger samples of component elements in a future application, that is, in this case larger numbers of items, classes, and students. A "what if?" analysis allows us to predict the likely effect of such increases on G coefficients and other parameters.

For example, all else being equal, we have the following:

- With 10 classes following each teaching method in place of five, remaining with 15 students per class, the G coefficients would increase in estimated value from 0.78 to 0.84 (relative measurement) and from 0.70 to 0.76 (absolute measurement).
- With 20 items in place of 10, the relative coefficient would increase to 0.81 and the absolute coefficient to 0.77.
- With 10 classes per method, still with 15 students per class, and with 20 items in place of 10, the relative coefficient would increase to 0.88 and the absolute coefficient to 0.83.

These are just three of the numerous optimization strategies that could be explored to increase the reliability of measurement to an acceptably high level (see Chapter 3 for further details).

Some points to note

From this brief overview of the methodology we would identify four points that are particularly worthy of note:

- We have highlighted the fact that while classical test theory is concerned with studying measurement reliability in the context of differentiation among individuals, G theory in addition enables us to evaluate the quality of measurement when we try to differentiate among things other than individuals, including subgroups. In this type of application the general logic of the approach is the same; what changes is the study objective and the way that different analysis designs are conceived.
- When designing an evaluation study, the objects to be measured must be clearly identified, as must those factors that might affect the reliability of the measurement results. Practitioners must be able correctly to identify the factors that are at play in their applications and to categorize them appropriately as objects of study or as contributors to measurement error.
- An important consequence of the points made above is that a measurement procedure is not necessarily a multifunctional tool. Thus, for example, a measurement procedure that is perfectly adequate for learning about the characteristics of individuals (attainment, motivation, and level of interest) can be totally inadequate for evaluating the difficulty of each of a set of test items comprising an examination paper, the progress made by a group of students between key points in a course, or the degree to which particular objectives are mastered by a population of students.
- Finally, in contrast with classical test theory, coefficient values, though useful as global indicators of measurement quality, are not the central focus of a generalizability study. The strength of the methodology lies in the detailed information it provides about the relative importance of the different sources of measurement error, information that is used to calculate appropriate standard errors of measurement and that can further be used to identify how the measurement procedure itself might be improved.

Content and objectives of this volume

Our principal aim in this volume is to illustrate the application versatility of G theory. In doing so, we overview the essentials of the theory with minimum reference to the mathematics that underpins it. The interested reader can readily study the theory itself in more depth by consulting, for example, Brennan (2001), Cardinet and Tourneur (1985), Cronbach et al. (1972), and Shavelson and Webb (1991). We rather describe the logic underpinning the G theory approach, and offer examples of application in different fields and contexts. We discuss the kinds of interpretation that G study analyses lend themselves to, showing in particular how analysis results can support a number of important decisions that might lead to improvement in the quality of the measurement tool or procedure used.

We begin by briefly but systematically introducing in Chapter 2 the terminology used in this domain and, most importantly, the basic concepts associated with the methodology. In Chapter 3, we offer a brief overview of the stages and procedures that are involved when carrying out a G study using the software package EduG, as well as guidance in the interpretation of results. EduG was developed by the Canadian company Educan Inc., with funding from a number of Swiss and Canadian academic and research institutions, which is why it is available in both English and French at http://www.irdp.ch/edumetrie/logiciels.htm.

We move on in Chapter 4 to describe and comment on a series of concrete examples of application in psychology and education. Readers who download EduG and the data sets relating to the application examples will be able to carry out the analyses themselves. Students looking for more practice in using EduG and interpreting its results will find further exercises in Chapter 5, with questions and commented answers. In Chapter 6, the final chapter, we review past stages in the development of G theory and reflect on some possible new directions for application. Finally, for the benefit of readers with a statistical background, a series of appendixes focus on technical matters, including the formulas used by EduG.

chapter two

Generalizability theory
Concepts and principles

The principal aim of a generalizability study, or G study, is to evaluate the characteristics of a given measurement procedure and to estimate measurement precision. To this end, the different sources of measurement error in play must first be identified, so that their relative importance as error contributors might be quantified. Relative importance is indicated by relative sizes of estimated variance components, typically estimated using classical ANOVA. It is these estimated components that are used to calculate measurement errors and G coefficients. It is these same components that are used in optimization studies, or "what if?" analyses, to predict how errors and coefficients might change in response to future changes in study design.

This chapter should be seen as an introduction to the concepts and terminology of G theory. Its mathematical content is minimal, in part because we assume that readers will have some prior familiarity with the ANOVA, on which G theory rests. Those who need to refresh their knowledge of ANOVA, or who would simply like to acquire an intuitive understanding of the technique, will find a very brief introduction in Appendix A. Readers interested in studying the mathematical foundation of G theory further should consult the seminal text produced by Cronbach and his colleagues (Cronbach et al., 1972), and the more recent publication by Robert Brennan (2001). For those who can read French, look also at Cardinet and Tourneur (1985). For a lighter overview of the theory, see Shavelson and Webb (1991).

G studies can be carried out on data that already exist (historic examination data, perhaps, or psychiatric patient records), or a study can be designed to provide data with a G study specifically in mind (marker reliability studies and program evaluations, for example). The difference is that in the former case some of the more important variables that might contribute to measurement error might be missing, in consequence limiting the potential value of the G study results, whereas in the latter case a careful study design can be employed to ensure that the important variables do feature. Either way there are clear steps to follow if a G study is to be successfully conducted. This chapter identifies those steps, and in so doing anticipates what users of the EduG software will need to think about when preparing for a G study analysis. In other words, the chapter is written with the needs of future G theory practitioners in mind, and focuses on:

- Data structure (Observation design)
- Facet-level sampling (Estimation design)
- Study focus (Measurement design)
- G coefficients (Design evaluation)
- The standard error of measurement (SEM) (Estimation precision)
- D studies (Optimization).

Chapter 3 explains how these steps are implemented with the help of EduG.

Data structure (observation design)

Facets and their inter-relationships

The quantitative data that we analyze (the measurements, typically scores) relate to the *dependent* variable in ANOVA terminology. In educational and psychological measurement contexts, this will be ability, skill, aptitude, attitude, and the like, with measurement taking the form of test scores, ratings, judgments, and so on. The *independent* variables, the factors, jointly comprise the qualitative conditions under which these data are obtained, some of which represent sources of measurement error. Examples include test items, occasions of testing, markers, raters, degree of topic exposure, among others. So where does the term "facet" come in? A facet in G theory is synonymous with a factor in ANOVA. The new term was originally introduced by Guttman, to avoid confusion in psychometric circles with the factors of factor analysis. Cronbach and his associates followed this lead in their early presentation of the developing G theory. In this volume we use the terms factor and facet interchangeably. When a single factor/facet is involved, its levels constitute the different conditions of measurement, for example the different test items within a mathematics test, or the different teachers marking the same student essay. When several facets feature, then the sources of variation that need to be taken into account will further include facet interactions, that is, interactions between the levels of one facet and those of another, such as markers differing in the severity or leniency with which they judge different student essays.

	Items				
	1	2	3	...	k
Students 1					
2					
...					
n					

Figure 2.1 Grid illustrating the crossing of Students with Items (SI).

When planning a G study, the first thing a researcher must do is identify the facets that are at play in the measurement process, along with their inter-relationships. This results in the "observation design." There are two different types of relationship: facets are either "crossed" with one another, or one is "nested" in the other. We say that two facets are "crossed" when every level of one of the facets is combined with every level of the other in a data set. Thus, if a *k*-item knowledge test is given to a group of students, and every student attempts every item, then the two facets Students and Items are crossed: this is symbolized as S × I, or more simply as SI. Think of the comparative study introduced in Chapter 1, where two methods of mathematics teaching were compared in terms of impact on students' mathematics interest. There, every student in the study completed the same attitude questionnaire, so that students were indeed crossed with items. We then have a score for every student on every item, which could be presented in the two-dimensional grid shown in Figure 2.1. The concept of crossing can naturally extend to cases where there are more than two facets. Had the comparative study involved the testing of students at the beginning and end of their programs, that is, if they had been tested on at least two different occasions, then students, items, and occasions of testing would be crossed one with another.

We say that two facets are "nested" if each level of one is associated with one and only one level of the other. There were 10 classes involved in the comparative study, with two mathematics teaching methods being evaluated. Five classes followed one teaching method and five followed the other, none following both, in parallel or consecutively. In that case, classes were nested within teaching methods, since any one class was associated with one only of the two teaching methods (Figure 2.2).

Methods	Method A					Method B				
Classes	1	2	3	4	5	6	7	8	9	10

Figure 2.2 Grid illustrating the nesting of Classes within teaching Methods (C:M).

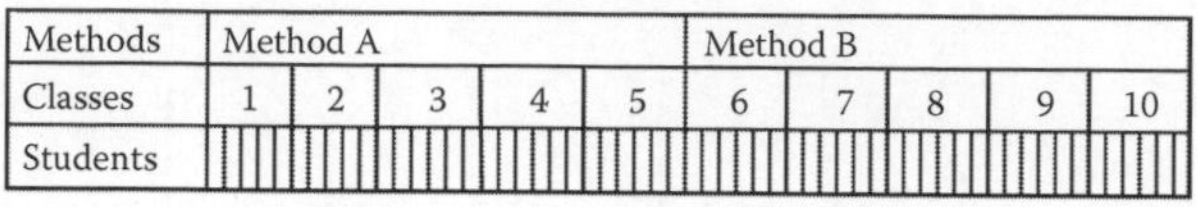

Methods	Method A					Method B				
Classes	1	2	3	4	5	6	7	8	9	10
Students										

Figure 2.3 Grid illustrating the nesting of Students in Classes and in teaching Methods (S:C:M).

More formally, the facet Classes was nested within the facet teaching Methods, a relationship symbolized as C(M), or C:M. As in crossing, nesting relationships can apply to more than two facets. The comparative study again offers an illustration, since students were nested in classes that were themselves nested within teaching methods (Figure 2.3). Notationally we represent this nesting hierarchy as S:C:M.

At this point, we note that when an analysis involves facet nesting, the analysis is much simpler if the number of levels of the nested facet is the same for every level of the nesting facet (the same number of students in every class, the same number of classes in every school, the same number of items representing each ability domain, and so on). Sometimes this requirement is not "naturally" satisfied, but it is usually possible to impose conformity retrospectively by relatively simple means, typically by randomly eliminating "excess" students from some classes (an analysis procedure that takes advantage of *all* the available data is presented in Appendix B for a simple design).

Variance partitioning

If we can identify all the various factors that potentially contribute to variation in a set of measurements, then we can "partition" the total variance to reflect these different variance sources. The aim of the ANOVA is then to assess the relative importance of the identified sources of (score) variation, through the estimation of variance components. The model for variance decomposition is represented in a score decomposition equation. For example, for the simplest two-facet crossed design *SR*, in which student essays (*S*) are independently marked by each of a number of different teacher raters (*R*), the score decomposition equation would be

$$Y_{ij} = \mu + S_i + R_j + SR_{ij}, e_{ij} \tag{2.1}$$

where Y_{ij} is the score obtained by student *i* from rater *j*, μ is the overall mean score (the average of all the individual student–rater scores), S_i is the effect of the student *i* (i.e., the deviation of the *i*th student's average score from the overall mean score), R_j is the effect of rater *j* (i.e., the deviation of the *j*th teacher rater's average assigned score from the overall mean score),

SR_{ij} represents the student–rater interaction effect (i.e., the degree to which rater j deviates from his/her normal level of severity/leniency in favor or otherwise of student i), while e_{ij} represents the error term that is confounded[i] with this interaction effect.

By assuming independence of effects, we can show through algebraic derivation that the total score variance (σ_T^2) can be represented as the sum of the three variance components: students (σ_S^2), raters (σ_R^2), and the student–rater interaction ($\sigma_{SR,e}^2$), this latter confounded with residual variance:

$$\sigma_T^2 = \sigma_S^2 + \sigma_R^2 + \sigma_{SR,e}^2.$$

(2.2)

A useful way of illustrating variance decomposition, and indeed of visualizing the observation design, is through a variance partition diagram, a vehicle originally introduced by Cronbach et al. (1972, p. 37). Figure 2.4 provides the variance partition diagram for the *SR* design discussed above.

Variance partition diagrams resemble classical Venn diagrams in that they comprise a series of intersecting circles (or ellipses). But they are conceptually different. In a Venn diagram each circle represents a given set of objects that satisfy some criterion, such as being algebra items. Intersections between circles represent a particular subset of objects—those that belong to each of the intersecting circles. Thus we might have a circle representing algebra items and another representing "problem solving" items. The intersection of the two circles would represent those problem solving items that tapped algebra skills. The interpretation of Figure 2.4 does not follow this same logic. We have two circles. But they do not represent sets of objects. Rather, the three areas marked out by the two circles represent the contribution to total score variance of three "effects" (in ANOVA

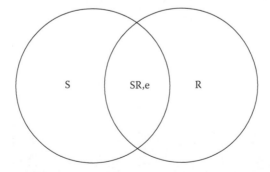

Figure 2.4 Variance partition diagram for the simplest model SR, where S and R represent Students and Raters, respectively, and both facets are random.

terminology). These are the Students effect, the Raters effect, and the Student–Rater interaction effect, the last of these confounded with all unidentified sources of systematic variance plus variance deriving from random fluctuation (hence the symbol "e"). These are the three identified constituents of the total variation in scores in this case.

This partition might be adequate if the students belonged to a single class and the teacher raters shared similar characteristics, although even then we might question whether there might be other potential sources of score variation that could and should be identified. Gender would be an obvious example of an effect that could be identified as a factor affecting students' scores. For their part, the teacher raters could perhaps also be classified as levels within a relevant descriptive category, such as gender again, or length of teaching experience, since these might affect their standards of judgment. In principle, the more facets we can identify and bring into the picture of variance partition, the better we will understand the nature of our measurements (provided only that we still have enough observations representing interaction effects to ensure stable estimates of variance components).

A different kind of variance partition diagram is necessary to represent facets that are nested one within the other. Nesting is symbolized in these diagrams by concentric circles (or other chosen shapes). Figure 2.5 illustrates the situation where students' essays are independently rated by a group of teachers, as before, but this time the students are identified as belonging to different classes (note that here and throughout the rest of this book we omit the "e" in the designation of the confounded highest order interaction effect). In place of the three distinct areas in Figure 2.4 we see in Figure 2.5 five different areas. These represent the five contributions to total score variation that are at play in this design: the

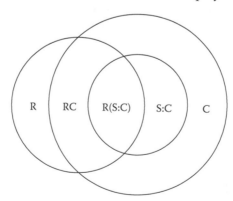

Figure 2.5 Variance partition diagram for the model R(S:C), where R, S, and C represent Raters, Students, and Classes, respectively, and all facets are random.

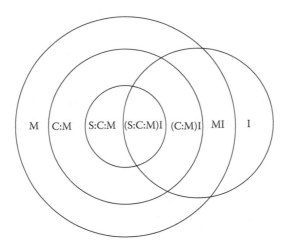

Figure 2.6 Variance partition diagram for the model (S:C:M)I, where S, C, M, and I represent Students, Classes, Methods, and Items, respectively, and all facets are random.

main effects for teacher Raters, Classes, and Students (within Classes), along with the interaction effects Raters–Classes and Raters–Students (this latter confounded with all unidentified systematic sources of variance and random effects).

As a third example, Figure 2.6 shows the variance partition diagram for the comparative study described in Chapter 1. In this study, five classes of students followed one teaching method, whereas five others followed an alternative teaching method, all students were given the same mathematics attitude questionnaire at the end of their programs, and 15 students represented each of the classes in the generalizability analysis. Here we have seven identified sources of variance: teaching Methods themselves, Classes (within methods), Students (within classes), Items, and the interaction effects Methods–Items, Classes–Items, and Students–Items, the latter as usual confounded with any residual variance.

ANOVA allows us to estimate the contributions to total score variation of the constituent sources shown in Figures 2.4 through 2.6, in the form of estimated variance components (see Appendix A, and Thompson, 2003, for computation illustrations).

Facet-level sampling (estimation design)

Fixed and random facets

There is an observation design associated with every data set, which, as just outlined, describes the data structure in terms of facets and their inter-relationships. But knowing the data structure is not in itself sufficient information for the correct estimation of variance components nor for defining the basic G theory model. There is another characteristic of facets that it is important to recognize: this is their sampling status. Facets can be termed "fixed," "random," or "finite random."

G theory applies sampling theory to the improvement of measurement procedures. The population sizes of many of the facets that feature in measurement applications are typically very large, and perhaps even infinite, at least from a practical point of view—Students and Items are examples—and we can only observe samples drawn from these populations. A facet is said to be random when the levels taken into consideration are randomly selected from the respective population or universe of interest: for example, 50 students from the population of fourth grade students, 20 items from among all those theoretically available for assessing a particular skill or ability, four schools from the 12 schools in a particular district, and so on. Sampling is said to be "purely random," or "infinite random," if applied to an infinite population. If samples are drawn randomly from finite populations then the facet's sampling status is "finite random."

A facet is said to be "fixed" when all its levels feature in the data set, no sampling of levels having occurred. "Natural" fixed facets are rare. Most of the time, they arise when a subset of population levels is explicitly identified for inclusion in a study. An example would be Districts, should all the sub-regions belonging to a given region be included in the study. Then no sampling occurs. A consequence is that any potential contribution to error variance from this variance source is eliminated. In such cases, the facet concerned is not intrinsically fixed, but is treated as such computationally. It is intended to make no contribution to measurement error. Its empirical mean is treated as if it were the true mean and its variance is taken as the average squared deviation from this mean.[ii]

Modifying a facet's sampling status is possible in both directions. Facets that might in principle be considered as random can be considered fixed, and vice versa, depending on the purpose of the study. In the words of Searle, Casella, and McCulloch (2006, p. 15): ". . . the situation to which a model applies is the deciding factor in determining whether effects are to be considered as fixed or random." For example, a particular subset of available levels can be deliberately included in a G study and others excluded, because these are the only levels of interest to the researcher. The selected subset might be the schools in one particular district, the objectives in a particular educational program, the beginning and end of a learning sequence, and so on. There will be no intention to generalize the results of the study to levels beyond those actually included in the analysis—for example generalization to other school districts, to other programs, to other time points in a learning sequence. The inverse is equally possible. For instance, the facet Moments (with two levels, "before" and "after"), although apparently fixed, is considered random for the computation of a *t*-test or an *F*-ratio. Only the interpretation of the between-level variance will be different. For a fixed facet, the variance component will measure the importance of the effect on the levels already observed. For a random facet, it will estimate the global importance of the expected source of variance in any repetition of the experiment, as the levels observed will be different each time (cf. Abdi, 1987, p. 109).

A measurement procedure can simultaneously involve both fixed and random facets, and random sampling can be carried out within a finite universe. Such models are termed "mixed." The sampling status of a facet has implications both for the algorithms used to estimate variance components and for the treatment of the facet in the computation of measurement error and G coefficients. Deciding the status of a facet when different status possibilities are available is therefore a critical step in defining the estimation design.

One of the difficulties associated with mixed models is that the variance is not calculated in the same way for fixed and purely random facets. In the latter case the mean square is calculated by dividing the sum of squares by the number of degrees of freedom, which is equal to the number of squared values minus 1. For a fixed facet, the sum of squares is divided by N, the number of levels in the facet population. The two types of computation do not follow the same logic: one gives an estimate of an unknown value, while the other reports an observed result. In consequence, the two types of estimation are essentially different and cannot be combined. To insure comparability of both types of variance estimates, a correction is possible, which involves multiplying the value produced by ANOVA by $(N-1)/N$, a coefficient that depends on N, the size of the population. The correction was proposed by Whimbey, Vaughan, and Tatsuoka (1967), who explain: "Since with a fixed-effects variable the J levels used in the experiment exhaust the whole population of levels, rather than being a sample of size J from a larger number of levels, J rather than J – 1 is the appropriate divisor" (of the sum of squares). The correction has no effect when population sizes tend to infinity, but for small population sizes it can make quite a difference. In the extreme case, when $J = 2$ the variance estimate obtained from a random sample is twice as large as when it comes from a fixed facet.

Graphical representation of the estimation design

In addition to illustrating the inter-relationships among facets, as in Figures 2.4 through 2.6, it is also useful in variance partition diagrams to

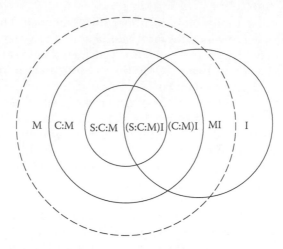

Figure 2.7 Variance partition diagram for the estimation design (S:C:M)I, where S, C, M, and I represent Students, Classes, Methods, and Items, respectively, and M is a fixed facet.

represent graphically the type of sampling that has been applied to each facet. We will see below that the composition of the error variance can be inferred (i) from the position of the areas representing each variance component in the diagram and (ii) from the type of sampling used for each facet: hence the usefulness of having both kinds of information displayed in the same diagram. The accepted convention for distinguishing between fixed and random facets diagrammatically is to represent purely random facets by continuous lines and fixed facets by dotted lines. Random finite facets may be symbolized by a combination of dots and strokes. The updated variance partition diagram for our comparative study (S:C:M)I is shown in Figure 2.7, in which the facet teaching Methods is represented by a dotted line to indicate its fixed nature.

Study focus (measurement design)

Relative and absolute measurement

The identification of (a) the facets that feature in the data set to be analyzed, (b) the nature of the facet inter-relationships, and (c) the sampling status of each facet is all the information needed for the ANOVA to be launched, and the respective variance components correctly estimated. The generalizability analysis proper begins at this point, using the estimated components in an appropriate way to provide information about measurement reliability and estimation precision. But a G study analysis can proceed only when two additional pieces of information are offered.

One is an identification of those facets that constitute the "object of study." Applying the principle of model symmetry to the simple model SR above, for example, gives two choices for the object of study: Student essays (S) or teacher Raters (R). The practitioner decides which.

The second piece of information that is in principle needed for a G study to commence is whether the measurement is to be "relative" or "absolute." The first is relevant when individuals or objects are measured and then positioned relative to one another within a distribution or classification. In this case, the measurement essentially consists of determining distances between entities ("There is a difference of 8 cm between the height of this individual and of that one."). This is a "relative measurement." The second consists of determining the exact position each individual occupies on the measurement scale ("This individual is 170 cm tall, the other is 178 cm."). The aim here is to identify the precise location of an individual or object on the respective measurement scale, quite independently of the other elements that might belong to the same distribution. This is "absolute measurement." Both types of measurement are affected by measurement error, with absolute measurement typically affected more than relative measurement.

With these additional vital pieces of information those variance components that contribute to "true score" variance and those that contribute to "error variance" can be identified, and estimation precision explored.

Differentiation, instrumentation, and generalization variance

"Differentiation variance" is synonymous with the concept of "true score" variance in classical test theory. It is that part of the total variance that "explains" the effective differences between measured entities, such as individuals or teaching methods. The differentiation variance is not directly observable, but is estimated by a single variance component or a linear combination of variance components. If Student essays (S) were the object of study in the SR model, then the differentiation variance would be simply the Students variance component. In the more complex comparative study, where the facet Methods is the object of study, the differentiation variance still comprises one single variance component: that for Methods. All other things being equal, the higher the differentiation variance, the more precise (reliable) will be the measurement, relative or absolute.

All conditions of observation serve as instruments for the measurement in question, and hence produce "instrumentation" variance. In the SR example the different teachers who rated the students' essays represent measurement conditions. Differences among them contribute to instrumentation variance. In the comparative study, both the students and the questionnaire items they respond to are the conditions of measurement, contributing instrumentation variance when Methods are being compared. Not all conditions of observation are sampled, however. Those that are fixed have no effect on the quality of measurement. They might be instrumentation facets, but they do not cause "generalization variance," the variance of interest in G theory.

"Generalization variance" is synonymous with error variance, or sampling variance. In other words, generalization variance is that part of the total variance that is attributable to fluctuations arising from the random selection of the components of the measurement procedure itself (students, classes, occasions of observation, and markers), as well as to purely random unidentified effects. In the case of relative measurement, generalization variance arises from interaction effects between differentiation facets and random instrumentation facets (i.e., generalization facets). In the case of absolute measurement, generalization variance is also produced by interactions between differentiation facets and generalization facets, but here the variances of the generalization facets themselves and of the interactions among generalization facets also contribute. All else being equal, the lower the generalization variance, the more precise (reliable) will be the measurement.

Faces of measurement

In a particular G study, the chosen objects of measurement determine the facet(s) of differentiation. There might be several such facets, if comparisons are to be made on several dimensions. For instance, if students are to be compared, irrespective of their gender, then the differentiation will combine the facet Gender with the facet Students (within Gender). Similarly, if the students' socio-economic origin also features in the study, then the differentiation will consider the students within each combination of gender and socio-economic origin: student differentiation will involve the facets Students (within Gender by Socio-economic origin), Gender, Socio-economic origin, and the interaction between Gender and Socio-economic origin. We call the set of all differentiation facets the "face of differentiation." Its levels are formed by the Cartesian product of the levels of each facet of differentiation. Symmetrically, the Cartesian product of the levels of all facets of instrumentation constitutes the set of all possible conditions of observation considered in the study. We call this the "face of instrumentation." Where all the instrumentation facets are random facets, the face of instrumentation will be synonymous with the face of generalization. But should any of the instrumentation facets be fixed, then the face of generalization will be a subset of the face of instrumentation, restricted to the set of sampled instrumentation facets. We use a forward slash, /, to distinguish those facets that form the face of differentiation from those that form the face of instrumentation. Thus, S/R indicates that in the SR model it is Student essays that are being differentiated with teacher Raters, the instrumentation facet.

Graphical representation of the measurement design

Variance attribution diagrams, which build on variance partition diagrams, are a useful graphical tool at this stage for indicating how the various effects are distributed among the three variance categories— differentiation, instrumentation, and generalization. Let us take the simple situation shown in Figure 2.4 as a first example. Suppose that our interest is in differentiating among students, that is, in spreading students on the measurement scale on the basis of their average essay marks, per- haps ranking them for some purpose. The differentiation facet in this case is Students, and it is the Students variance alone that comprises the differentiation variance.

But which variance sources contribute to error, or generalization, vari- ance? For relative measurement the answer is the Student–Rater interaction variance, divided by the number of teacher raters that will have contributed to each essay's average rating, that is, σ_{SR}^2/n_r. This is because the presence of this variance would indicate that one teacher's ranking of essays would differ from another teacher's ranking, thus contributing to uncertainty in the overall result. Figure 2.8 illustrates this, vertical hatching indicating the differentiation, or "true," variance and crossed hatching indicating the interaction variance, source of generalization, or "error," variance.

If, alternatively, we were interested in estimating students' actual ("absolute") achievement scores for essay writing, rather than their achievement relative to that of their peers, then the magnitude of the

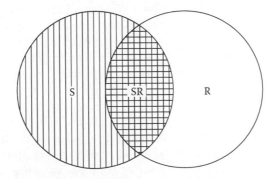

Figure 2.8 Variance attribution diagram for relative measurement for the design S/R, where S and R represent Students and Raters, respectively, and both S and R are random facets.

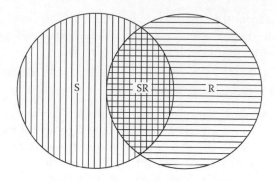

Figure 2.9 Variance attribution diagram for absolute measurement for the design S/R, where S and R represent Students and Raters, respectively, and both S and R are random facets.

marks given to the essays by the teacher raters would become very important. If the teachers involved in the essay rating could have been substituted by others, then Raters would be considered a random facet, and differences in overall judging standards would now also come into play as a source of measurement error. The main effect of a generalization facet (here the between-rater variance) is symbolized by horizontal hatching. The Raters facet would join the Student–Rater interaction as a source of random fluctuation (Figure 2.9), the two variances jointly representing "absolute" generalization variance. The generalization variance itself is calculated by adding the two components, after first dividing each one by the number of teachers involved [$(\sigma_R^2/n_r) + (\sigma_{SR}^2/n_r)$], to reflect the fact that the students' average achievement scores would be the result of this number of independent "observations."

If, on the other hand, the teachers were considered to be the only raters of interest, then they would represent themselves alone, and not any larger group of teacher raters who could have done this same job. In this case the facet Raters would be fixed, and it would not contribute to any kind of measurement error. Student measurement, relative or absolute, would in principle be error-free. However, there could be no generalization of results beyond the particular group of teacher raters involved.

We note that while a fixed instrumentation facet, like Raters in this example, has no impact on the precision of relative or absolute measurement, it nevertheless still exists as a source of variance. The facet will be included in the ANOVA table as a source of score variance, along with its interactions with other facets. But the sum total of its effects for each object of study will always be null, by the very definition of fixed effects in the model. Consequently such facets contribute only to "passive" variance.

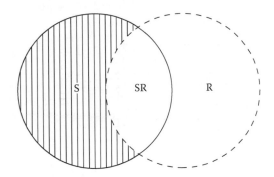

Figure 2.10 Variance attribution diagram for relative and absolute measurement for the design S/R, where S and R represent Students and Raters, respectively, and R is a fixed instrumentation facet.

This is why they are represented as "empty" areas in variance attribution diagrams, such as that shown in Figure 2.10.

We have so far discussed the measurement of students in this example. But as we have noted earlier, we could equally have been using the essay rating design to compare the rating behavior of teachers as raters. The focus of the measurement would then shift from students to raters, with the facet Raters becoming the differentiation facet in place of the facet Students. Assuming that the facet Students is random (i.e., that the students involved can be considered as a representative sample of all similar students, rather than as important individuals in their own right) the Student–Rater interaction would remain a component of generalization variance, contributing to error variance for both relative and absolute rater measurement, this time divided by the number of student essays rated by each teacher, that is, σ_{SR}^2 /n_s. The between-student variance would be a second contributor to generalization variance in the case of absolute rater measurement (Figure 2.11), again divided by the number of student essays involved, that is, σ_S^2 /n_s.

But what of the comparative study? The object of study in this case was teaching methods, and Methods is the differentiation facet. If the classes that followed one or other teaching method were considered samples representing all similar classes, and if the students in the classes were also considered samples representing all similar students, then both Classes and Students, as random facets, would contribute to generalization variance for both relative and absolute measurement. This is because different classes and different students might well have produced a different result. If the facet Items is also considered a random facet, then the interactions between this facet and Methods, Classes, and Students will also contribute to generalization variance for relative measurement (Figure 2.12).

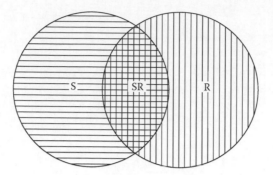

Figure 2.11 Variance attribution diagram for absolute measurement for the design R/S, where S and R represent Students and Raters, respectively, and S is a random instrumentation facet.

In each case the variance component contributing to generalization variance will be divided by the number of observations that produced it (the relevant sample size) in the error variance expression. Thus with Classes, Students, and Items as random facets, the expression for the generalization, or error, variance, with Methods the object of study, will be

$$\text{Relative error variance} = \sigma^2(\delta) = \frac{\sigma^2_{\text{C:M}}}{n_c} + \frac{\sigma^2_{\text{S:C:M}}}{n_s n_c} + \frac{\sigma^2_{\text{MI}}}{n_i} + \frac{\sigma^2_{\text{(C:M)I}}}{n_c n_i} + \frac{\sigma^2_{\text{(S:C:M)I}}}{n_s n_c n_i}$$

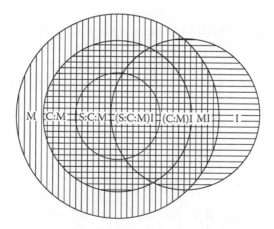

Figure 2.12 Variance attribution diagram for relative measurement for the measurement design M/SCI, where M, S, C, and I represent teaching Methods, Students, Classes and Items, respectively, and all facets are random.

For absolute measurement the Items variance would further increase the generalization variance, as follows:

$$\text{Absolute error variance} = \sigma^2(\Delta) = \frac{\sigma^2_{C:M}}{n_c} + \frac{\sigma^2_{S:C:M}}{n_s n_c} + \frac{\sigma^2_{MI}}{n_i} + \frac{\sigma^2_{(C:M)I}}{n_c n_i}$$

$$+ \frac{\sigma^2_{(S:C:M)I}}{n_s n_c n_i} + \frac{\sigma^2_I}{n_i}$$

It might be useful at this point to offer some general guidelines about how the components of absolute and relative error variance can be properly identified in variance attribution diagrams, and how the components of the differentiation variance can be located:

1. To identify the components of absolute error variance, hatch horizontally the circles representing generalization facets. (Disregard the circles representing fixed instrumentation facets, which should be left blank.)
2. To identify the components of relative error variance, hatch vertically the areas already hatched horizontally that lie within the circles representing the differentiation facet(s).
3. To identify the components of the differentiation variance, hatch vertically the areas within the differentiation facet circles that are outside of the instrumentation facet circles.

These rules are important because they give the key to the computation algorithm used by EduG. They show that the attribution of each source of variance to the differentiation or generalization variance obeys three simple rules, independently of the particular design considered in the G study.

G coefficients (design evaluation)

Reliability coefficient

A reliability coefficient summarizes the results of a G study, because it indicates the degree to which the measurement instrument or procedure used is able to differentiate reliably among the individuals or objects (such as teaching methods) concerned. In other words, it tells us whether the results produced are satisfactorily dependable, regardless of the specific components that define the particular instrument or procedure (this set of items rather than some other, for example). In this case, we consider that the results obtained do not depend, or depend in negligible proportions, on the particular subset of elements one had available (the 20 items comprising a test, for example, or the five classes following one or other teaching method). We would say, therefore, that the results are generalizable to the population (finite or infinite) of all the elements that could have been used to develop the measurement instrument (universe of admissible conditions of observation). Such a conclusion is considered reasonable if the reliability coefficient is equal to at least 0.80.

It is convenient to compute, for any measurement design, a generic reliability coefficient that we call Coef_G. It is a ratio of estimated differentiation variance to estimated total variance of observed scores:

$$\text{Coef\_G} = \frac{\hat{\sigma}_D^2}{\hat{\sigma}_D^2 + \hat{\sigma}_G^{2'}}$$

where $\hat{\sigma}_D^2$ and $\hat{\sigma}_G^2$ represent, respectively, the estimated differentiation variance and the estimated generalization variance (relative or absolute, as appropriate). The value of Coef_G will depend on the specific nature of the measurement situation.

The differentiation variance is calculated simply by summing the estimated variance components for those facets and facet interactions that constitute the face of differentiation. As shown earlier, the generalization variance is calculated by summing the estimated variance components for those facets and facet interactions that constitute the face of generalization, after first dividing each component by the appropriate facet sample sizes, that is, by the number of sampled levels representing each generalization facet in the data set.

Three types of G coefficient

Depending on the type of differentiation planned, three coefficients can be calculated: a coefficient of relative measurement, a coefficient of absolute measurement, and a coefficient of criterion-referenced measurement.

Loosely formulated, classical G theory focuses on the proportion of total score variance that can be attributed to the "true" variation among randomly sampled objects of study. This proportion is indicated in the "coefficient of relative measurement." This is the coefficient that Cronbach et al. (1972) defined specifically as the "G" coefficient, and symbolized by $E\hat{\rho}^2$ (expected rho squared). It is an indicator appropriate to relative measurement. This coefficient enables us to estimate how precisely the procedure can locate individuals or objects of measurement relative to one another, and to estimate correctly the distances between them—for example, a distance of 3 or 5 points between the score of one individual and the score of some other. (For a simple crossed design, like the SR design discussed earlier, it corresponds to the α coefficient of classical test theory.)

The "coefficient of absolute measurement," on the other hand, that Brennan and Kane (1977a,b; see also Brennan, 2001, p. 35) defined as the "dependability" coefficient, Φ, evaluates the ability of a procedure to locate individuals or objects reliably on a scale in *absolute* terms. The formula is identical with that of $E\hat{\rho}^2$, the difference between the coefficients lying in the composition of the generalization variance, Φ typically having more contributions to this variance than $E\hat{\rho}^2$. For instance, Φ takes into account the variation in question difficulty, and not only the interaction students × questions. For this same reason, while the values of $E\hat{\rho}^2$ and Φ can sometimes coincide, most typically Φ is lower in value than $E\hat{\rho}^2$. Because we use Coef_G as a generic term to designate all the ratios of estimated true variance to estimated total score variance, we distinguish $E\hat{\rho}^2$ from Φ by speaking of relative and absolute Coef_G.

Developed at the same time as Φ by Brennan and Kane (1977a,b), $\Phi(\lambda)$, the "coefficient of criterion-referenced measurement," extends Φ to address cut-score applications (Brennan, 2001, p. 48). This coefficient indicates how reliably an instrument can locate individual results with respect to a threshold (or cut-off or criterion) score set at point λ on the measurement scale. For example, if the cut-score for a test was set at 60 points on a 0–100 scale ($\lambda = 60$), $\Phi(60)$ would indicate how reliably the test could place individual students on one or other side of this point (estimation of the distance between the individual scores and the chosen cut-score). Starting from calculation of the coefficient of absolute measurement, the criterion coefficient takes account not only of the variability of the individual scores around their mean, but also of the distance between this mean and the cut-score (the differentiation variance is in this case a combination of these two elements).[iii]

The three types of G coefficient just presented have in common the fact that they have all been developed for infinite facets sampled randomly (like students or items). In the following section we look at reliability coefficients from a quite different perspective.

Special case of a fixed differentiation facet

When a differentiation facet is fixed, then the usual relative coefficient of reliability $E\hat{\rho}^2$ is no longer appropriate. It is replaced by ω^2. Statisticians (for instance Winer, Brown, & Michels, 1991) distinguish between these two situations—random differentiation facets versus fixed differentiation facets—because the variance between objects of measurement for a fixed facet is directly observable, whereas it has to be estimated for an infinite random facet. The two models are clearly different. And since the formulas used to calculate $E\hat{\rho}^2$ and ω^2 are different, the values obtained are not the same, ω^2 generally being lower in value than $E\hat{\rho}^2$, or at best equal.

This theoretical hurdle is an inconvenience, because it is clear that the two coefficients derive from the same logic. In nontechnical terms, both indicate the proportion of true score variance between objects of study that is contained in the total score variance. Looking more closely at what it is that distinguishes the computations of the coefficients ω^2 and $E\hat{\rho}^2$, we see that it is simply the fact that with a fixed facet the between-level variance is calculated by dividing the sum of squares by N (the number of squares summed, equivalent both to the sample and to the population size), whereas with an infinite random facet it is the mean square that provides an unbiased estimate of population variance, so that the divisor of the sum of squares is $n-1$ (with n the number of levels sampled). If Whimbey's correction for the size of the universe (Whimbey et al., 1967) is applied, as mentioned earlier in regard to the mixed model, the estimated variance components become comparable, all then accepting the classical definition of variance as the mean squared deviation. At this point, the same formula for Coef_G can be applied to all facets within the framework of G theory, regardless of the method of sampling.

Appendix C is devoted to an algebraic proof of the equivalence between Coef_G, ω^2, and $E\hat{\rho}^2$ under the assumption of a classical definition of variance. Derivations are carried out for both a crossed and a nested design. Coef_G, like the two other indicators, represents the proportion of estimated true variance in the estimated total score variance, but Coef_G has the advantage that it is equally applicable to the intermediate case of finite random facets. Coef_G goes even further, by embracing the case of absolute measurement when both ω^2 and $E\hat{\rho}^2$ refer only to relative measurement. Coef_G may represent the proportion of universe ("true") variance in the total variance, when that total variance includes the absolute error variance. The wider applicability of Coef_G can be presented another way. Brennan and Kane (1977a) have introduced Φ as a reliability coefficient for absolute measures, but Φ and $E\hat{\rho}^2$ are applicable only to faces of differentiation that are entirely random. Coef_G is the only absolute reliability coefficient applicable to fixed or finite facets of differentiation. Thus, Φ, $E\hat{\rho}^2$, and ω^2 can all be considered particular cases of the generic reliability coefficient Coef_G.

Care should be taken to apply correctly the customary rules concerning interactions with fixed facets. In most measurement applications, especially in education and psychology, the objects of study are typically drawn at random, while some conditions of observation might be fixed. Interactions between random differentiation facets and fixed instrumentation facets do not contribute to measurement error, and constitute only passive variance. But this is no longer true when the roles are reversed.

When a differentiation facet is fixed, its interactions with random instrumentation facets (generalization facets) do contribute to error, because the effects of these interactions on each object of measurement no longer add up to zero in the model. As a mnemonic rule, one can remember that a dotted line may be disregarded when entering the blank area of a fixed instrumentation facet, but not when passing the border in the other direction, from a differentiation facet to a generalization facet (none of which is blank in that case). The risks of making such errors with fixed differentiation facets are minimized when EduG is used, thanks to the algorithms underpinning the computations. But the downside is a higher chance of misunderstanding the nature of the SEM applicable to fixed differentiation facets. This we attempt to address in the following section.

Standard error of measurement (estimation precision)

The SEM—the square root of the measurement error variance—is a critical piece of information to consider when evaluating the measurement qualities of an instrument or procedure. Its use has a long history in the field of psychometrics, and is recommended in the Joint Standards: "The standard error of measurement is generally more relevant than the reliability coefficient once a measurement procedure has been adopted and interpretation of scores has become the user's primary concern" (AERA, 1999, p. 29). *The Publication Manual of the American Psychological Association (APA)* similarly declares: "The reporting of confidence intervals (for estimates of parameters, for functions of parameters such as differences in means, and for effect sizes) can be an extremely effective way of reporting results" (APA, 2001, p. 22).

The calculation of the SEM represented for Lee Cronbach the essential contribution of G theory, as he confirmed, with the help of Shavelson (2004), in his final thoughts on the 50th anniversary of the alpha coefficient (Cronbach & Shavelson, 2004). When interpreting the output of a G study, it is the SEM that informs the user about the size of error affecting the results in the context of relative or absolute measurement. It provides concrete and directly interpretable information about the reliability of the results produced by the measurement procedure: for example, it has a probable error of 0.05 on a 0–1 scale, or of 7.8 on a scale running from 0 to 50, and so on. Moreover, if we assume errors to be Normally distributed, we can conclude that in around two-thirds of cases (specifically 68%) the actual error will be lower than the standard error, and that in the remaining one-third or so of cases it will be higher. The SEM further allows us to define margins (or zones) of uncertainty, in particular the usual confidence intervals associated with each particular measurement. The SEM can be considered as the standard deviation of the (theoretical) distribution of all the errors that will arise should different "equivalent" samples be interchanged.

In some measurement contexts the SEM is the only relevant parameter to estimate. Sample-based attainment surveys represent an important example. In such surveys, not only are schools and students typically sampled, but so too are test items. Moreover, the test items used in large-scale surveys are often distributed among several different test booklets, so that booklets can also be considered randomly sampled. Matrix sampling is usually used to distribute booklets randomly among students. The principal aim in such surveys is to estimate the average performance of a given population of pupils on a defined domain of test items, for example the average numeracy performance of 12-year-olds. A subsidiary, and equally important, aim is to estimate the SEM attached to the population performance estimate.

How can we understand the SEM in practice? The SEM of a measuring instrument can be imagined as its inherent arbitrary "fluctuation," that adds an amount of error in one direction or the other to the measure it records for each "object of study." We cannot know the "true" value of what we are measuring, but repeating the process for the same object might eventually provide a good approximation to it, in the form of the mean of the various observations. In this way we can estimate the size of error in each case. The SEM, as the square root of the variance of these errors, quantifies the precision, or imprecision, of the measuring procedure. It gives, in the units of the measurement scale itself, a sort of mean error, a "standard" error. In a testing context we do not, of course, expect the same student repeatedly to attempt the same test question. Instead we invite the student, and others, to respond to each of a sample of n questions, all focused on the same general aspect. In a simple nested design the error will be estimated as the deviation of each question score from the student's mean score. In ANOVA terms, the error variance will be the within-student variance. In a crossed design the error variance will be the variance of the interaction between Students and Items, obtained by subtracting from the total sum of squares the between-student and the between-item sums of squares, and dividing by the residual number of degrees of freedom. For both designs, the SEM will be the square root of the error variance, either "relative" or "absolute," depending on the nature of the measurement.

It is important to note that the SEM concerns a single object of measurement at a time and is not influenced by differences between them (between students in this example). When the "face of differentiation" is the Cartesian product of several facets, the set of all the objects of measurement is defined as all the possible combinations of the levels of one facet with the levels of the other facets. Differences between all these objects of measurement create the differentiation variance. Similarly, all the facets of instrumentation form, by the Cartesian product of their levels, their own total set of admissible conditions. The observed mean scores for one object of study in all these admissible conditions have a certain distribution. The SEM is the standard deviation of this distribution, but averaged for all objects of study.

In practice, the SEM is computed on the basis of the ANOVA-derived components of variance. Two cases are possible: the differentiation facet is defined as random, or it is defined as fixed. The first case is the conventional and familiar one, and presents no problem. This estimate tells us the size of the error that will affect each object of study if we repeat the measurement under the same conditions of observation (the universe of generalization). The computed G parameters provide us with the details. We can see what the contributions to measurement error of all relevant sources of variation are, as well as the size of the error variance itself, along with its square root, the SEM. But if the differentiation facet is defined as fixed, the Whimbey correction will have been applied to the components of variance in order to make them additive. The value of the SEM will be affected by this. In consequence, we will have two different values for the SEM, one computed with uncorrected components as usual, and the other with the variance components reduced to take into account the known size of their facet populations.

There is, in fact, no contradiction here. The two SEM provide different kinds of information. The reduced SEM informs us retrospectively while the other informs us prospectively. The "corrected" version measures the error variance that was affecting each object of study at the moment of the G study. This variance will be smaller than the variance indicated by the other "uncorrected" SEM and this is what it ought to be. Indeed, the computation of the expected error variances takes into account the uncertainty due to our ignorance of the true mean and for that reason tends to inflate the estimates somewhat. Thus both SEM give us some important information to consider separately and to apply in a suitable way.

Chapter 3 will explain how the two kinds of SEM are computed. EduG offers the corrected (reduced) SEM if the differentiation facet was defined as fixed. The uncorrected alternative (giving the expected error variance) can be obtained simply by re-defining the size of the differentiation facet as infinite, in which case random model components are computed and no Whimbey's correction is applied to them during the estimation of the error variances associated with the differentiation facet and its interactions.

D studies (optimization)

As we have already mentioned several times, the aim of a G study is to evaluate the characteristics of a measurement procedure, to identify its strengths as well as its weaknesses, so that constructive modifications and adjustments can then be made. The phase that aims at improving the procedure on the basis of an analysis of its characteristics is termed a Decision study, or D study. This involves defining an optimization design. This final step uses the results obtained in the G study, in particular the estimated variance components for the principal contributors to measurement error. Here, we perform "what if?" analyses, to explore and evaluate the different improvements that could be made by changing the measurement procedure: for example, increasing the number of items, students or classes, eliminating one or more levels of a facet, fixing a facet that was initially considered random, and so on.

Changing the number of facet levels sampled

A frequently used optimization procedure is to increase the number of levels sampled to represent a generalization facet. For example, the 10 items used in the comparative study might be increased to 20, or 10 classes could replace the previous five following one or other teaching method, and so on. Everything else being equal, this kind of adjustment usually serves to reduce the contribution to measurement error of the facet concerned and of its interactions, and thus increases measurement reliability. This procedure applies the same rule as the classical Spearman–Brown prophecy formula[iv] when only the Items facet is concerned. When several instrumentation facets feature, an optimization strategy can sometimes be identified that involves increasing the number of levels for facets with large error contributions while decreasing the representation of facets with low impact on measurement error, in order to keep the total number of observations approximately the same. The generalizability of the mean score for each object of measurement can thus be increased without necessarily increasing the cost of the examination, a possibility that did not exist before G theory was developed.

Interestingly, in certain cases a form of optimization can be achieved by reducing rather than increasing the number of sampled levels of a facet. A 50- or 80-item test, for example, might prove too long to be reliable, because of the effects of insufficient testing time, student fatigue, concentration lapses, and so on. Analysis sometimes shows that reducing the number of items by a third or a quarter has a minimal effect on measurement reliability. In this case, a slight drop in reliability would be adequately compensated by noticeable improvements in applicability.

Eliminating atypical levels for certain facets (G-facets analysis)

One particular approach that sometimes features at this stage of the process is G-facets analysis, which can be useful when the number of levels of a facet is small, for checking whether any particular levels have a disproportionate influence on the instability or imprecision of the measurement. A particular "mathematics interest" item might, for example, be answered in a certain way relatively more frequently by girls than by boys, or by pupils in one class compared with those in others (which could be considered evidence of "bias," or might possibly indicate a flaw of some kind in the item). To rectify "problems" of this type, that is, the presence of interaction effects, which in the case of test items might indicate item bias, the atypical level(s) can be eliminated from the data set. This strategy, though, should always be adopted with great caution, since it can reduce the validity of the measurement tool or procedure, by changing the nature of the universe of generalization. It should be used only when theoretical and/or technical reasons justify it, for example if it is clear that a test item is poor for some observable reason or that the class in question had been subject to a very different curriculum experience than the others.

Changing the number and nature of facets

Sometimes there can be high variability associated with the levels of a generalization facet because they represent different dimensions or belong to distinctly different domains. Such a situation can arise when, for example, the items in a knowledge test jointly evaluate several objectives or relate to various topics in a teaching program. It is always possible to group the items and to nest them within a new, fixed, facet (items nested within objectives or within program topics, in this case). This strategy will serve to reduce the measurement error to some extent, since it will now be influenced only by the between-item variance within each level of the nesting facet and not by the nesting facet itself, since fixed facets make no independent contribution to error variance. Another strategy that can be applied in certain situations is to "fix" a facet that was originally considered random. For example, in the comparative study, where the original intention was to generalize results beyond the particular classes involved (the facet Classes being therefore defined as random), error variance could be reduced by newly defining the Classes facet as fixed, and accepting the consequent loss of power of generalization.

In both these cases, but especially in the second, it must always be recognized that the strategy adopted carries important implications for the inferences that can be drawn from the analysis results. This is mainly because it changes (reduces) the breadth of the universe of generalization— the universe to which the reliability of the measurement tool/procedure can be extended.

Eliminating measurement bias[v]

The three optimization strategies just described manipulate instrumentation facets. Alternative strategies focus rather on differentiation facets. For illustration, suppose that we want to differentiate among the students in the different classes following one particular teaching method on the basis of their responses to the questionnaire items. The facet Classes, which by default will be a differentiation facet (as a nesting facet for Students), could be a source of measurement error if between-class variability is high. In effect, since the objective is to differentiate among students, a high between-class variation would suggest that belonging to one class rather than another will have an important influence on students' results. In consequence, the observed deviations between students will be confounded with differences between classes, something that could lead to an incorrect and misleading interpretation of the results. To avoid this type of problem, the analysis could be repeated after replacing the original student scores with difference scores, the differences being between the students' scores and their relevant class averages. To estimate the likely effect on reliability of this modification, it would be sufficient to eliminate the between-class variance component from the Coef_G numerator.

chapter three

Using EduG
The generalizability theory software

EduG is unique among other available generalizability software packages in that it was conceived specifically to exploit the symmetry property of G theory. It offers flexibility in the choice of object of study and identification of instrumentation facets. It also provides G coefficients that are appropriate in terms of the particular sampling status of the facets involved. EduG is Windows-based, user-friendly, readily obtainable, and free (downloadable from http://www.irdp.ch/edumetrie/englishprogram.htm). The use of the software for carrying out a G study follows the same logic as outlined in Chapter 2. Data entry is straightforward and the Workscreen easy to use. Usage, nevertheless, presupposes familiarity on the part of the user with the terminology and basic concepts of G theory. In particular, in order to use EduG it is essential to know how to define correctly the different designs that guide a G study analysis: that is, observation design, estimation design, and measurement design.

In the observation design, the facets represented in the data set are identified, along with their crossing and nesting relationships and observed levels. In the estimation design, the sampling status of each facet is defined as infinite random, finite random, or fixed, by indicating the number of levels in its universe. This information enables the software to partition the total variance appropriately in the ANOVA, and to estimate variance components correctly. Finally, the measurement design identifies those facets that together comprise the object of measurement—the differentiation face—and those that are part of the measurement procedure—the instrumentation face.

To illustrate the use of EduG, we look again at the comparative study introduced in Chapter 1 and considered further in Chapter 2. This features four facets: Methods, Classes, Students, and Items. Methods, M, is a fixed facet with two levels—specifically two different methods of teaching mathematics, Method A and Method B. Classes, nested within Methods, C:M, is an infinite random facet with five levels (i.e., classes) per method. Students, nested within Classes (and hence also within Methods), S:C:M, is similarly an infinite random facet, with 15 levels (i.e., students) per class. Finally, Items, I, is also an infinite random facet, with 10 levels—specifically 10 questionnaire statements evoking evidence of positive or negative attitudes to mathematics and mathematics learning.

The aim of the study was to evaluate the ability of the measuring instrument—the questionnaire—to distinguish reliably between the two teaching methods. This aim is important to bear in mind, given the discussion about actual and expected error estimates offered in Chapter 2. Where reliability is the focus, we are interested in estimating the proportion of true between-Methods variance in total score variance, which Coef_G will provide. If, on the other hand, we are interested in testing the statistical significance of the size of any observed difference between Methods, then we need prospective variance estimates that can validly underpin applied significance tests. We can use EduG to provide these, but actually for significance testing applications the use of more specialized software is advised.

We address the following points in this chapter:

- Opening main menu and Workscreen
- Specifying observation, estimation, and measurement designs
- Creating or importing a data file
- Requesting a G study
- Interpreting EduG reports
- Following with a D study.

Opening main menu and Workscreen

When called, EduG displays four main menu choices: *File, Edit, Preferences,* and *Help*. The *File* menu offers three options: *New, Open,* and *Quit*. As would be expected, these allow the user, respectively, to create a new "basis" (an EduG file containing data description, design details, and data), to open an existing basis, or to close the program. Through *Preferences* various program parameters can be defined, and a preferred text editor identified: results reports can be edited either in text format, the default, or in rtf format, involving a call to MS Word (screen displays and reports in text format use a fixed character font while in rtf, or rich text format, MS Word manages results presentations). Chosen parameters can be changed at any later point, before an analysis is launched, by modifying the options displayed on the left, in the middle of the Workscreen (Figure 3.1). The *Edit* option allows a call to whichever choice of text editor the user will have made under *Preferences*. Finally, the *Help* option offers a global overview of the software. It also allows consultation of the Index, through which numerous different help topics can be accessed as the user actually runs the program (more detailed information on this subject is given via an in-built *Help* facility).

EduG "basis" files are characterized by the extension "gen." They contain the data to be analyzed in a G study, a description of the structure of that data (observation design), and instructions for analysis (estimation and measurement designs). To create a basis the user selects the menu options *File/New*, then offers a file name for the new basis, and clicks "Open," at which point the Workscreen shown as Figure 3.1 appears.

A basis with the given file name is automatically created as soon as the Workscreen appears, albeit empty of any content. The next step is to offer design information.

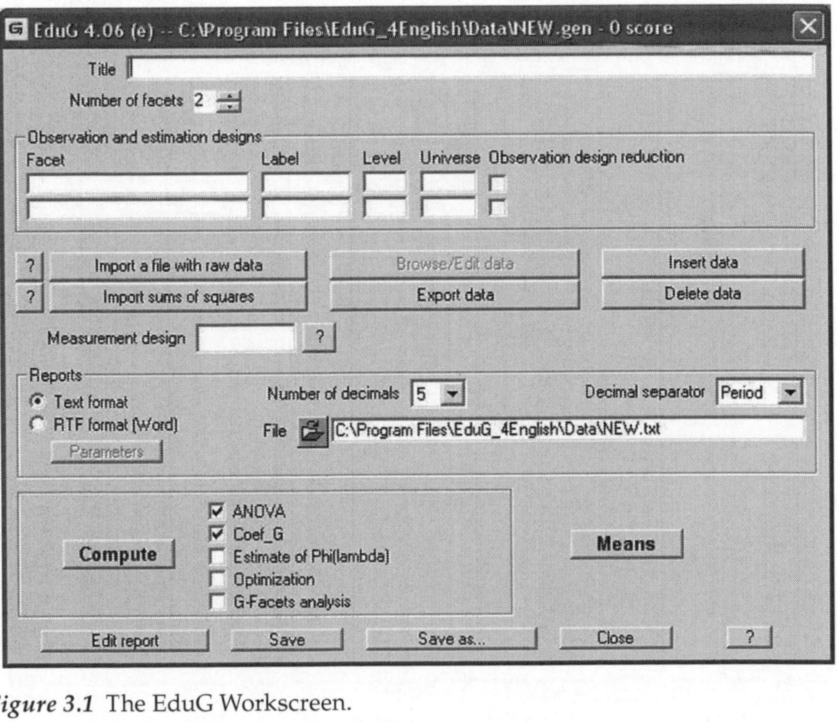

Figure 3.1 The EduG Workscreen.

Specifying observation, estimation, and measurement designs

Observation design

After having given an optional title to the basis at the top of the Workscreen, the user indicates the number of facets involved in the study, and specifies the observation and estimation designs (in the area highlighted in Figure 3.2 and expanded in Figure 3.3).

In principle, any given data set should be maximally exploited at this stage. This is to say that as many facets as possible should be identified for exploration in the analysis, within the constraints of data balance (equal cell sizes), data quantity (too few observations for a facet or facet interaction will lead to unstable estimation), and software limitations (EduG can handle up to eight facets in any one G study analysis). In our example, for instance, it would have been possible to identify each student as male or female, possibly as being socially deprived or not, as being a high achiever in mathematics or not, and so on. But with only 15 students representing

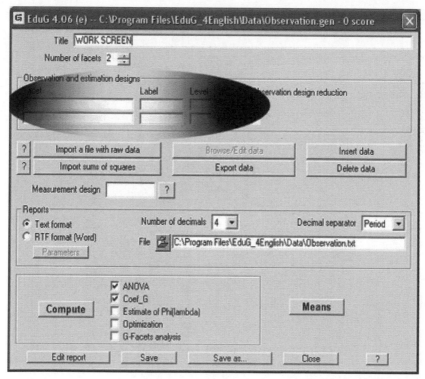

Figure 3.2 Area for specifying the observation and estimation designs.

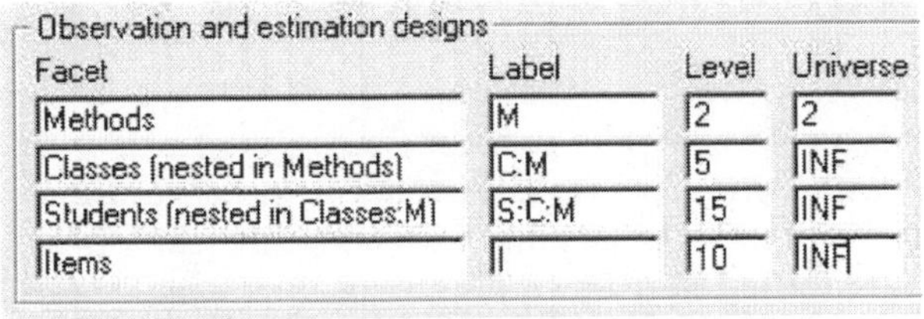

Observation and estimation designs			
Facet	Label	Level	Universe
Methods	M	2	2
Classes [nested in Methods]	C:M	5	INF
Students [nested in Classes:M]	S:C:M	15	INF
Items	I	10	INF

Figure 3.3 Declaring facets in the Workscreen.

each class within each teaching method, any further subcategorization would have resulted in at best single-digit representation in each cross-classification.

Each facet to be included in analyses is identified by its full name (e.g., "Methods") in the first column headed "Facet" in the "Observation and estimation designs" area (see Figure 3.3). This field can also be used to mention certain facet characteristics (e.g., "1st year students") or to record any other comments. The second column identifies each facet by a label, which for convenience would usually be its initial letter, for example, M for Methods and I for Items. Nesting relationships are the only ones that need be explicitly indicated, with the usual colon convention, for example, C:M in our example, for Classes nested within Methods, and S:C:M, for Students nested within Classes within Methods. EduG assumes all other facets to be crossed with one another, for example, M with I, and S (within C:M) by I. The number of levels representing each facet in the data set is given in the third column: "Levels." For instance, facet M has two levels only, representing the two teaching methods being compared, facet C has five levels (five classes per method), facet S has 15 levels, the 15 students in one or other of the 10 classes, and facet I has 10 levels, for the 10 items included in the attitude questionnaire.

This set of information—facet identification and numbers of levels observed—comprises the observation design, describing the structure of the data. Once defined, an observation design cannot be changed within an existing basis, although it can be temporarily reduced if an analysis is required that involves a subset only of the levels of a declared facet. The column "Observation design reduction" (right-most column in the facet declaration block shown in Figure 3.1) allows the user to exclude a part of the data set from analysis at any point, for example the data relating to teaching Method B if a G study analysis is required for Method A only (see later under "Requesting and interpreting a G study"). But should an analysis be required that would use the same data set with a differently defined structure—fewer or more facets, perhaps, or changed facet relationships—then a new basis must be created.

Estimation design

The size of each facet universe is given in the "Universe" column. For fixed facets the universe size will be the same as the number of levels included in the data set, as is the case for facet M; for infinite random facets the single letter I, for INF or infinite, suffices to describe the facet's infinite universe, as shown for facets S and I. Finite random facets will have universe sizes somewhere between the number of observed levels and infinity. It is the combination of "Levels" and "Universe" that defines the sampling status of each facet, and hence constitutes the estimation design. Where there is a choice of sampling status, it is for the user to decide the appropriate choice in terms of the degree of generalization desired. For instance, here we are treating Items as an infinite random facet, which means that the G study results will be generalizable to all items of similar type, and not just to those comprising the questionnaire. An alternative choice would be to consider Items as a fixed facet, in which case measurement precision would be higher, since fixed instrumentation facets make no contribution to measurement error. But then there would be no possibility of validly generalizing results beyond the set of items actually used.

Unlike observation designs, estimation designs can be modified, by changing facet universe sizes, without the need to create a new basis. This is useful for a model with a fixed differentiation facet since it offers a ready way of reanalyzing the data, by re-declaring the facet as random in order to compute an uncorrected SEM, should this be needed. For example, if you want to estimate the Methods effect in the comparative study, then you should be computing ω^2 as a measure of the effect of this factor. You will get this value as a relative Coef_G with Methods declared as a fixed differentiation facet with a universe comprising just two levels. Coef_G relative is then 0.78. But if you want to estimate the SEM for a future application of this design, you declare the facet as infinite and you obtain a relative SEM of 0.1009, with which you can compute a confidence interval for the difference score between the two methods.

Measurement design

The area for defining the measurement design appears on the center left of the Workscreen, as shown in Figure 3.4. It distinguishes those facets that together comprise the object of measurement, that is, the differentiation facets, from those that condition the measurement procedure, that is, the instrumentation facets. Concretely, in a measurement design representation, differentiation and instrumentation facets are distinguished by placement of their identifiers to the left and right of a slash: differentiation facets to the left and instrumentation facets to the right. In our example,

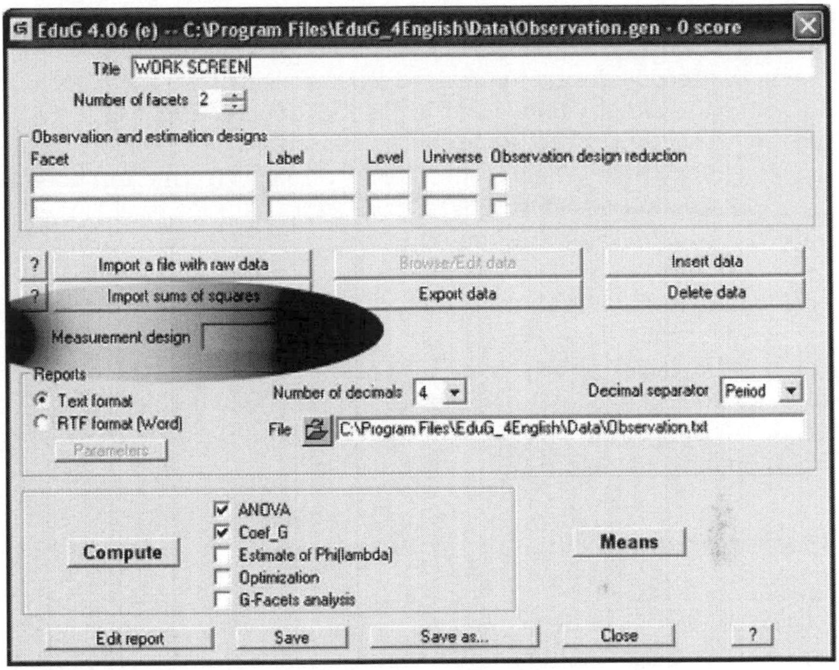

Figure 3.4 Area for declaring the measurement design.

M/CSI describes the design in which there is a single differentiation facet, teaching Methods (M), and three instrumentation facets, Classes (C), Students (S), and Items (I).

The location of the various facet identifiers in the expression is sufficient information for a measurement design. There is no need to indicate crossing or nesting relationships here, since this information is given in the observation design. So that although Students are nested within Methods in our example, we do not have to indicate this in the measurement design expression. An important rule that must be followed when specifying a measurement design, however, is that when a nested facet appears on the face of differentiation, then the facet(s) in which it is nested must appear there also.[i] Thus, for example, if the objective in our application had been to differentiate among students rather than teaching methods, then since the students are nested within classes and teaching methods, S, C, and M would appear to the left of the slash in the measurement design, which would then be MSC/I. This rule is not reciprocal, as we have seen: if we have a nesting facet (such as M) on the face of differentiation (the objective of the study being to differentiate among methods) it does not follow that the facet(s) nested within it (here C and S) should also be there—the nested facets in this case are properly on the face of instrumentation, giving the measurement design M/SCI.

A measurement design that has no differentiation facet is indicated for EduG purposes as a slash followed by the identifiers of *all* the facets in the design, whether they contribute to error variance or not. Thus, for a survey design involving high schools (H), students (S), items (I), and test booklets (B), the observation design might be (S:H) × (I:B), all four facets might be considered random (probably, though not necessarily all, with infinite universes), and the measurement design would be written as /HSIB in EduG's measurement design window. In such applications all random facets and their interactions will usually contribute to measurement error, that error being absolute measurement error. Like estimation designs, measurement designs can be changed without the need to create a new basis. The object of study might, for example, be changed from Methods to Items, and a new G study analysis requested using the same basis.

Creating or importing a data file

The data processed in a G study are assumed to be quantitative (metric), that is, expressed in terms of interval or ratio scales. However, as is often the case in the social and behavioral sciences, the data we have available are not always truly quantitative, at least to the extent that most physical measurements are (length, weight, speed, etc.). In some cases we work rather with ordinal variables, notably when we use knowledge tests or different types of psychological scale (of interest, motivation, depression, etc., such as that in the illustrative example featured here). Experience shows, however, that under certain conditions such variables can be treated as though they *are* quantitative. As a result, the data can be processed in a G study just as they can be subjected to other statistical techniques originally developed for metric variables (e.g., factor analysis or ANOVA). One important data constraint is that when measurements are made using several different items (or under several different conditions, in several situations, at different moments, etc.), they must be based on the same type and length of scale (0–1, 1–4, 0–10, or whatever). Sometimes this requirement is not satisfied initially: for example when some items are dichotomously marked, while others have maximum marks of 2, 5, and so on. In this type of situation, it is usually possible to impose scale uniformity after the event, by applying different types of score conversion or mapping. The most frequent strategy is to transform all item scores onto a proportional scale, which can be done by dividing individual item scores by the maximum score for the item (thus a score of 2 for an item with maximum score 5 would be equal to 0.4 on the transformed proportional scale).

If the *Browse/Edit data* button (in the Workscreen shown in Figure 3.1) is active, this shows that data have already been entered. It is then possible to inspect them, and if necessary, to correct these data. If this button is not active, then the *Insert* button next to it is active and gives the possibility of introducing either raw scores or sums of squares. This last possibility will be justified first.

EduG can only handle complete and balanced data. For example, in a set of item response data for the method comparison, there must be no missing responses (completeness), and item results must be available for the same number of students in each teaching method (balance). But practitioners sometimes find that their data are neither complete nor balanced. Incompleteness typically results from an individual omitting to respond to a test item or to an attitude statement in a questionnaire. Different strategies can be employed to address missing values, using criteria appropriate to the particular context. In educational testing contexts, for example, missing item responses are often simply replaced by zero scores, on the assumption that students skipping items would not have produced correct responses had they attempted them (this strategy would be inappropriate, however, should there be many skipped items toward the end of a test, which would suggest test fatigue rather than an inability to respond correctly). In other cases, regression analysis, or some other appropriate method, might be used to provide estimated item scores. In every case, what is important is that the problem of missing values must be resolved before the data set is offered to EduG for analysis.

Unbalanced data can arise in two main ways: as a result of an inherently unbalanced design, for example where different numbers of test items are used to represent different curriculum domains (number skills versus fractions, perhaps), or because of data loss, for example where students who were intended to be tested were absent for the testing session. In the first case, appropriate weighting formulas can be used to take advantage of all the data at hand. An example is offered in Appendix B. In the case of data loss, as we have already indicated, this type of problem can be resolved by randomly discarding "excess" observations (facet levels) from the data set, such as items or students.

In many cases, however, replacing missing data points is difficult to do when no obvious criteria for replacement exist. An alternative solution that can be employed is to use EduG to compute variance components starting from sums of squares rather than from raw scores, using some other software package to calculate the sums of squares. This option is also useful in situations where there is no access to the raw data, but ANOVA results are available (in published research reports, for example). Introducing sums of squares to EduG via the keyboard is the procedure that requires the least typing effort, since the names of the facets, their

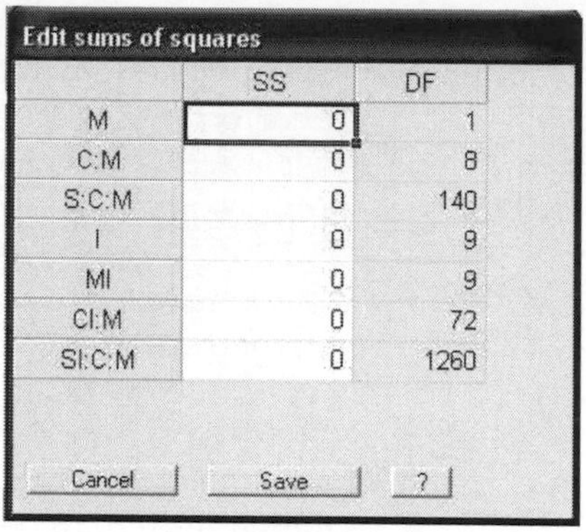

Figure 3.5 Introducing sums of squares via the keyboard.

inter-relationships, and their degrees of freedom are already known to the software, on the basis of the declared observation design. The facet labels and the degrees of freedom presented by the software serve to guide appropriate entry of the sums of squares (Figure 3.5).

Entering the same information by using a file import is almost as simple: select *Import sums of squares* in the Workscreen, and then identify the name of the data file to be imported. The data file must have been prepared beforehand in text format, each row containing a facet label, its sum of squares, and degrees of freedom, separated by a semicolon. The rows can be introduced in any order, as EduG rearranges the information appropriately to match the order in which facets appear in the declared observation design. For a simple crossed design comprising four rows and five columns, the file content might look like this:

R;10;3
C;16;4
RC;30;12.

If the data to insert are raw scores, EduG would display a table like that in Figure 3.6, in which the order of columns follows the order in which the facets were defined in the observation design (for our example, this is M, then C:M, then S:C:M, then I). Raw scores are entered, as appropriate, in the final column of the table: here, the first raw score to be entered would be that associated with teaching method 1 (Method A), class number 1 within this method, student number 1 within this class, and item 1. When data entry (along with inevitable corrections) is completed, the resulting data set must be saved.

Often a data file already exists that contains the data of interest in a suitably tabulated form, created using some other software. Such files can be imported directly into EduG, after conversion to ASCII format, by clicking

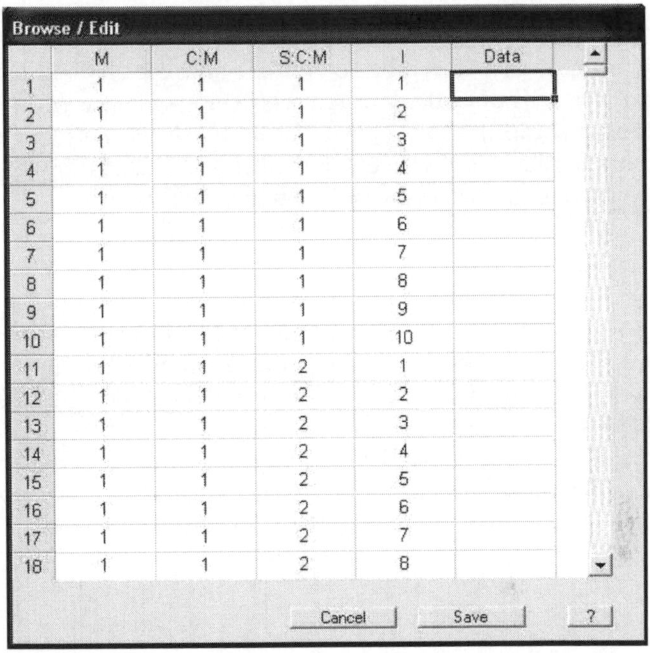

	M	C:M	S:C:M	I	Data	
1	1	1	1	1		
2	1	1	1	2		
3	1	1	1	3		
4	1	1	1	4		
5	1	1	1	5		
6	1	1	1	6		
7	1	1	1	7		
8	1	1	1	8		
9	1	1	1	9		
10	1	1	1	10		
11	1	1	2	1		
12	1	1	2	2		
13	1	1	2	3		
14	1	1	2	4		
15	1	1	2	5		
16	1	1	2	6		
17	1	1	2	7		
18	1	1	2	8		

Figure 3.6 Introducing raw scores via the keyboard.

on *Import a file with raw data*. For a file import to be successful, care must be taken that the order in which facets are declared in the observation design conforms to the structure of the data file to be imported. The first facet to be declared is the one whose levels change least rapidly when the file is scanned from left to right and row by row (the first sort variable). The last facet to be declared would be the one whose levels change fastest (the last sort variable). EduG confirms, or not, that an import is successful and indicates the number of records now in the basis. Given the problems that can arise during a file import, it is recommended that even when a successful import is confirmed the imported data be visually scanned, to check that they have indeed been properly associated with the levels of the various facets. A click on the *Browse/Edit data* button will give access to the data.

One final situation that can sometimes arise is when the file for import comprises a single column of data rather than a grid, that is, all the scores are recorded in some predefined order within a single column. A file of this type can also be directly imported. But, once again, attention must be given to the order in which facets are declared in the observation design, to ensure a correct reading of the data (appropriate matching of scores to the level combinations for the different facets). We note finally that further information about data capture is to be found in EduG's in-built help facility.

Requesting a G study

Several different analyses can be requested. These include a full G study analysis, with or without calculation of a criterion-referenced coefficient, a D study (optimization) analysis, a G-facets analysis, and calculation of simple means for each of the levels of the different facets as well as of the level combinations for two or more facets in interaction.

Means

The request procedure for the calculation of level means is very simple. After having defined the observation and estimation designs and prepared, directly or through import, the target data file, the user simply clicks on *Means* (lower right of the Workscreen shown in Figure 3.1). EduG responds by listing the relevant facets and facet interactions, so that the user can identify those for which means are required. There is no need for EduG to access a measurement design for this particular analysis.

G study analysis

To carry out a G study analysis, an appropriate measurement design should have been identified, in addition to the observation and estimation designs. For the comparative study, the measurement design is identified as M/CSI and is entered into the Measurement design window as described earlier. We repeat that since in this particular design the differentiation facet Methods is fixed, the associated SEM computed by EduG will not be appropriate for inferential tests. Thus, we would not recommend using EduG with a design like this for any purpose other than estimating measurement reliability. To do this, the boxes *ANOVA* and *Coef_G* should be checked and *Compute* clicked (lower left of the Workscreen in Figure 3.7). In studies where a criterion-referenced dependability coefficient $\Phi(\lambda)$ is also needed, then the relevant choice box for this computation must also be checked, and the cut-score, λ, given (the score separating successful test-takers from others). Where it would be meaningful to use the Optimization facility or a G-facets analysis, their choice boxes, too, should be checked (Figure 3.8). To request an observation design reduction, it suffices to check the Observation design reduction box associated with the facet of interest. EduG lists the levels of the relevant facet, inviting the user to identify those levels to be ignored in a G study analysis (e.g., level 2 of the Methods facet if an analysis is required for Method A only). The numerical labels relating to the levels to be excluded are automatically displayed alongside the checked reduction box, as shown in Figure 3.9.

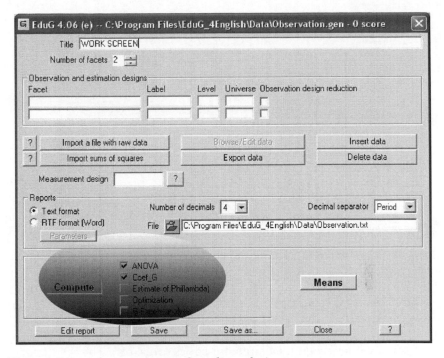

Figure 3.7 Area for requesting a G study analysis.

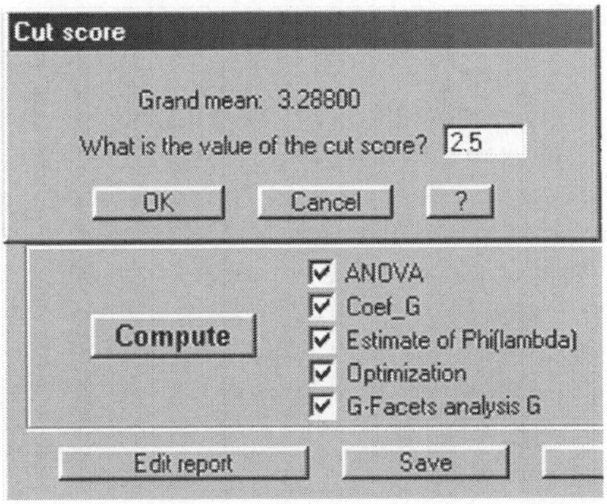

Figure 3.8 Requesting analyses through the Workscreen and notably an estimate of Φ(λ).

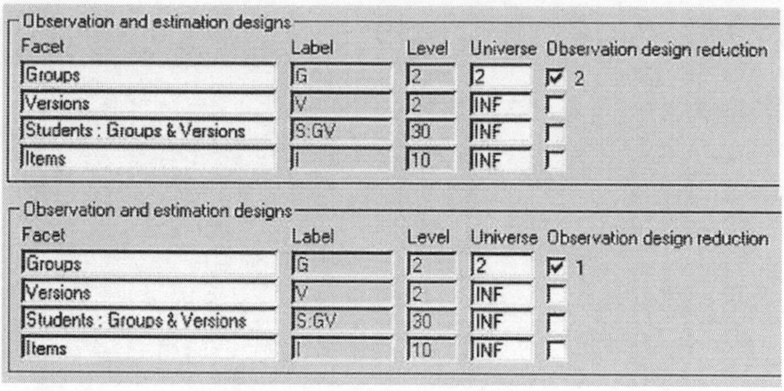

Figure 3.9 Requesting an observation design reduction.

Interpreting EduG reports

Current report parameters are indicated in the center of the Workscreen, and can be changed before any analyses are requested: output file name, file format (text or rtf), destination directory (using the "Save as" button), choice of decimal separator and number of decimal places in table displays (the default is five—be careful to retain an appropriate number of decimal places, since variance components and error variances can have very small values when measurement scales are short). The results produced by EduG comprise a standard set of tables, to which may be added other supplementary information relating to any optional analyses the user might have requested. The simplest one is the computation of means, but results can also be printed for a criterion-referenced coefficient, an optimization or a G-facets analysis.

Means

Values of the means can be obtained for all possible distinct sources of variance: for each level of any facet and for level interactions across facets. For our example, lists will be produced giving the mean for each method, for each class in each method, for each item, for each item for each method, and so on. As the number of possible means can be very large in some designs only those specifically requested are printed. A variance is also printed alongside each mean; this is the observed (not the expected) variance of values for those elements that have contributed to the mean.

ANOVA table

The standard output begins with a summary of the underlying observation and estimation designs (in the comparative study, all facets are random, except Methods, which is fixed). This is followed by the ANOVA table, as shown in Table 3.1 for the teaching methods comparison with no observation design reduction. For each facet or facet interaction the table gives the usual sums of squares, degrees of freedom, and mean squares. In addition, three sets of estimated variance components appear: one set relating to an entirely random effects model, one set relating to a mixed model, and a final set of components resulting from application of Whimbey's correction factor (see Chapter 2). The penultimate column shows the percentage contribution of each source to the total variance (sum of the corrected variance components). A final column presents the standard error associated with each (random model) variance component. Note the relatively high contribution to total score variance of the facet Students, along with the interaction effect between Students and Items (this is confounded with all unidentified sources of variance). The variance contributions of the Methods effect and of the interaction between Methods and Items are low.

The variance component estimates in Table 3.1 all have positive values. In theory, negative variances are not possible, of course. But in practice sampling fluctuations around small component values sometimes do result in small negative component estimates. Where negative component estimates emerge in an analysis, these appear in the ANOVA table. But, conventionally, they are replaced by zeros in the follow-on computations of generalizability parameters. Other variance components have necessarily null values, such as those derived from a one-level facet. An example would be an analysis for one gender only, after an observation design reduction in which the other gender was temporarily excluded. In such cases a row of dots represents the null values in the ANOVA table, and

Table 3.1 ANOVA Table for the Comparison of Teaching Methods

Source	SS	df	MS	Random	Mixed	Corrected	%	SE
M	48.600	1	48.60	0.055	0.055	0.027	6.4	0.053
C:M	33.690	8	4.211	0.017	0.017	0.017	4.0	0.013
S:C:M	156.233	140	1.116	0.092	0.092	0.092	21.4	0.013
I	55.560	9	6.173	0.013	0.036	0.036	8.5	0.021
MI	37.440	9	4.160	0.046	0.046	0.023	5.3	0.024
CI:M	52.785	72	0.733	0.036	0.036	0.036	8.3	0.008
SI:C:M	248.667	1260	0.197	0.197	0.197	0.197	46.1	0.008
Total	632.975	1499					100%	

again a zero value is carried through ensuing calculations. No null value appears in the current example.

G study table

The next table to be presented is the G study results table itself (Table 3.2). In this table, the various facets and their interactions are separated into differentiation facets (the first two columns) and instrumentation facets. Here also, the different error variances are displayed. These are based on the data shown in the ANOVA table (Table 3.1), and are calculated using the estimation algorithms appropriate to the sampling status of the different facets. For example, the contribution of students to relative error variance, namely, 0.0012, is simply the corrected variance component for students (0.0919, from Table 3.1) divided by the total number of observed levels of this facet sampled for each object of measurement. Here the divisor is the number of students representing each teaching method in the data set (75, given 15 students in each of five classes).

The final "% abs." column indicates how the absolute error variance is apportioned among the different contributory sources. This information allows us to identify those sources of variance that have the greatest negative effect on the precision of the measurements, information that would be useful in a follow-on D study analysis to show how measurement precision might be increased. In this example it is clear that the facets Classes and Items are the largest contributors to measurement error (for "absolute" Method measurement), together contributing over 60%, followed by the Methods-by-Items interaction effect, accounting for another 20%, and Students, accounting for over 10% more. The other two sources of variation—interaction between Classes and Items and between Students (within Methods) and Items—play a less important role, their cumulated effect resulting in less than 10% of the total absolute error variance. It is increasing the numbers of classes (and hence also of students) and/or the number of items in a future application that would serve to increase measurement precision.

In the penultimate two rows of Table 3.2, the total differentiation variance and the total generalization variance (here synonymous with instrumentation variance) are shown, along with the standard deviations (or standard errors, calculated in the usual way by finding the square root of the corresponding variances). The differentiation variance is more than three times larger than the relative error variance (at 0.0273 and 0.0079, respectively) and more than twice as large as the absolute error variance (0.0115). With the differentiation and error variances, relative or absolute, one can determine whether a weak G coefficient is attributable to high measurement error, or to minimal differences between the objects measured

Table 3.2 G Study Table for the Teaching Methods Example
(Measurement Design M/SCI)

Source of variance	Differentiation variance	Source of variance	Relative error variance	% Relative	Absolute error variance	% Absolute
M	0.0273		—		—	
	—	C:M	0.0034	43.2	0.0034	29.6
	—	S:C:M	0.0012	15.5	0.0012	10.6
	—	I	—		0.0036	31.5
	—	Ml	0.0023	28.9	0.0023	19.8
	—	CI:M	0.0007	0.9	0.0007	6.2
	—	SI:C:M	0.0003	3.3	0.0003	2.3
Sum of variances	0.0273		0.0079	100%	0.0115	100%
Standard deviation	0.1653		Relative SE: 0.0889		Absolute SE: 0.1074	
Coef_G relative	0.78					
Coef_G absolute	0.70					

(in this regard, remember that for a given error variance the value of the G coefficient increases or decreases with the differentiation variance).

The two G coefficients—relative and absolute—are displayed at the bottom of the G study table. These offer a global indication of the reliability of the measurement procedure, and give a general indication of the precision of the measurements produced. The values of the two G coefficients, 0.78 for relative measurement and 0.70 for absolute measurement, certainly suggest that measurement precision with the current numbers of classes, students, and items is not entirely satisfactory. As noted in Chapter 1, the results obtained here indicate that the research study was not really capable of providing a sufficiently precise measure of average student interest for each of the two teaching methods, nor of the difference between them. At the end of the G study report EduG records the overall mean, along with its sampling variance and standard error, taking account of the estimation design. (Remember that the facet Methods has been defined as fixed, thus lowering Coef_G and the SEM.)

Standard error of measurement

More detailed analyses can then proceed. The way the SEM is computed can be inferred from the G study results table (Table 3.2). The components of "true score" variance are identified in the first column, with their respective contributions alongside, in the second column. In the third column are listed all other components, whether or not they contribute to error variance. The fourth and fifth columns show the actual and relative (percentage) contributions of these components to relative error variance, while the sixth and seventh columns do the same for absolute error variance. In these columns, the interactions of a differentiation facet with a fixed instrumentation facet are put aside: they appear between parentheses and do not contribute to either error variance. The same is true of the components estimated as negative in the ANOVA computations. The various error contributions are summed, with the results shown at the bottom of the columns. The square roots of the error variances, the relative and the absolute SEM, respectively, appear one row below.

The SEMs have values 0.0889 and 0.1074, respectively, for relative and absolute measurement. But to produce confidence intervals around the estimates of average student interest level for each method, and around the estimated difference between the two teaching methods, the SEM to use[ii] is the one obtained with Methods declared infinite random, notably because in this case Whimbey's correction factor will be equal to 1 (and thus have no effect) for all interactions between the differentiation facet and the instrumentation facets. (All these interactions are sources of error and their variance estimates should not be reduced by correction factors.) The new values of the SEM, relative and absolute, are 0.1009 and 0.1074 (the absolute error variance remains stable because of compensating changes in the estimation of the variances).

The means for the two teaching methods, Method A and Method B, were, respectively, 2.74 and 2.38. The estimated difference in method means is therefore 0.36 on the 1–4 scale. Since the variance of a difference between two independent variables[iii] is the sum of their separate variances, or twice their pooled variance (as in this case), the variance of the difference in means between Method A and Method B is 2×0.1009, or 0.2018, giving a standard error of the difference of 0.4492. The 95% confidence interval around the difference in method means (1.96 times the standard error) is therefore ±0.880, giving an interval of 0.000–1.240 for the difference in method means. The average interest of students following one method or the other is estimated with an SEM for absolute measurement equal to 0.1074, quite near the SEM for relative comparisons. This gives a 95% confidence interval of ±0.211 around each method mean. For Method A, the interval will be from 2.53 to 2.95; for Method B, from 2.17 to 2.59.

A design with no differentiation facet

To illustrate an analysis for an application in which there is no differentiation facet, let us suppose that these two teaching methods are the only

Table 3.3 ANOVA Table for a G Study with No Differentiation Facet

Source	SS	df	MS	Random	Mixed	Corrected	%
				Components			
M	48.6000	1	48.6000	0.0546	0.0546	0.0546	12.0
C:M	33.6900	8	4.2113	0.0171	0.0171	0.0171	3.7
S:C:M	156.2333	140	1.1160	0.0919	0.0919	0.0919	20.2
I	55.5600	9	6.1733	0.0134	0.0134	0.0134	2.9
MI	37.4400	9	4.1600	0.0457	0.0457	0.0457	10.0
CI:M	52.7850	72	0.7331	0.0357	0.0357	0.0357	7.8
SI:C:M	248.6667	1260	0.1974	0.1974	0.1974	0.1974	43.3
Total	632.9750	1499					100%

two being used in a particular country, and that the purpose of the data collection was to estimate the mean level of mathematical ability of the population of students from which those who were tested were drawn. The focus of the G study would then be quite different, the aim now being to estimate an appropriate margin of error for the estimated population mean score. The results produced by EduG take the form shown in Table 3.3 for the first ANOVA.

There are two reasons for declaring Methods as infinite random here. Firstly, this G study aims at determining margins of error for the observed population mean, and we need the SEM estimating the expected error variance. Secondly, since G coefficients are not relevant, and cannot even be computed, in this type of application, there would be no value in setting up the data to be able to directly compare various contributions to total variance. That is why in Table 3.4, the G study results table, there is no "object of study," and no differentiation variance shown in the leftmost columns. Neither is there any entry in the "Relative error variance" column. Any error contributions that there are in this application are contributions to absolute measurement error, and these are shown in the "Absolute error variance" column, their relative percentage contributions appearing alongside. As usual, this information can be used in a follow-on D study to guide the choice of alternative sampling strategies to reduce the overall measurement error in the estimation of the population mean.

Clearly, no G coefficient can meaningfully be computed in this type of application, since there will be no differentiation variance to constitute the numerator in the variance ratio. In its place, EduG computes a coefficient of control. This is again a proportion, taking a value in the conventional 0–1 range. It is defined as the proportion of the total absolute error variance of the grand mean that is due to the main factors. These factors

Table 3.4 A Different G Study with No Facet of Differentiation (Measurement Design /MSCI)

Source of variance	Differentiation variance	Source of variance	Relative error variance	% Relative	Absolute error variance	% Absolute
	—	M	—		0.0273	80.9
	—	C:M	—		0.0017	5.1
	—	S:C:M	—		0.0006	1.8
	—	I	—		0.0013	4.0
	—	MI	—		0.0023	6.8
	—	CI:M	—		0.0004	1.1
	—	SI:C:M	—		0.0001	0.4
Sum of variances					0.0337	100%
Standard deviation			Absolute SE: 0.1837			

are those over which the experimenter has some control, because their sampled levels are explicitly determined. The complement to 1 for this coefficient of control is the proportion of absolute error variance that is due to interactions between the main factors. These interactions are difficult to control, because they are partially or wholly unpredictable. A high coefficient of control indicates that the sources of variance affecting the grand mean have a simple structure, relatively easy to describe and to understand.

Following with a D study

After reviewing the initial results of a G study analysis, users often want to carry out a further analysis with the aim of improving the qualities of the measuring instrument. EduG offers two possibilities, Optimization and G-facets analysis.

Optimization table

The first of these options, Optimization, offers the "what if?" analysis mentioned in Chapters 1 and 2. This allows users to see the likely effect on the relative and absolute coefficients of changing the number of levels of one or more facets, and/or of changing facet characteristics (e.g., "fixing" a facet that was initially considered random, or modifying the size of a finite universe). Such modifications are indicated in a table such as that

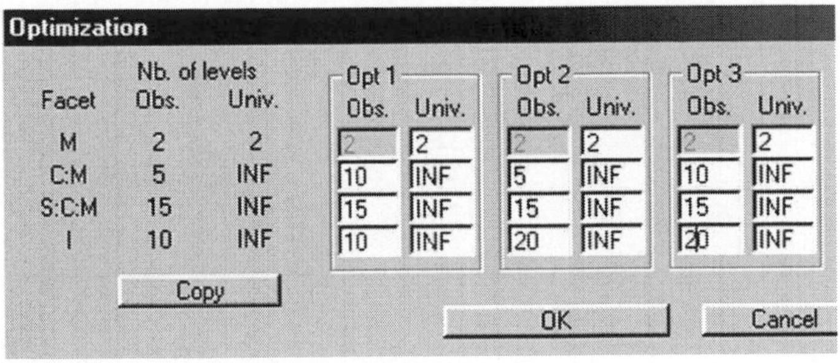

Figure 3.10 The optimization command window.

shown in Figure 3.10, which appears when the Optimization choice box is checked (bottom of the Workscreen—see Figures 3.1 and 3.8).

The observation and estimation design details are reproduced in the left-hand side of the optimization command window. On the right-hand side the user can indicate alternative numbers of observed levels of any generalization facet and/or an alternative sampling status (by changing the size of the universe). Note that the "Observed" cells associated with the differentiation facet M are shaded out, since no changes here would be meaningful.[iv]

Reverting to our original study in which the relative effectiveness of the two teaching methods was being explored, we see in Figure 3.10 a number of requested D studies. In Option 1, the number of classes following each teaching method is doubled, from 5 to 10, leading to a simultaneous increase in the overall number of students involved. In Option 2, the number of items is increased from 10 to 20, with class and student numbers kept the same. In Option 3, both item numbers and class numbers are simultaneously increased, from 10 to 20 in the first case and from 5 to 10 in the second. EduG responds to this optimization request by re-calculating the new contributions to measurement error of the various facets, and producing estimated G coefficients and other parameters for the alternative study designs (Table 3.5).

From Table 3.5, we see that increasing the numbers of items and/or classes involved in a repeat study would reduce error variance and so increase reliability for both types of measurement, relative and absolute. Simultaneously increasing the levels of both facets, Items *and* Classes, would raise reliability to an acceptable level, even for absolute measurement: the relative coefficient would increase to 0.88 from the original 0.78, and the absolute coefficient would increase to 0.83 from the original 0.70. Note, though, that these results were obtained on the assumption that the Methods facet was fixed. As mentioned earlier, this is fine as far as the interpretation and use of Coef_G is concerned. But while the reported

Table 3.5 Optimization Results for the Comparison of Teaching Methods

	G study		Option 1		Option 2		Option 3	
	Level	Universe	Level	Universe	Level	Universe	Level	Universe
M	2	2	2	2	2	2	2	2
C:M	5	INF	10	INF	5	INF	10	INF
S:C:M	15	INF	15	INF	15	INF	15	INF
I	10	INF	10	INF	20	INF	20	INF
Number of observations	1500		3000		3000		6000	
Coef_G relative	0.7756		0.8428		0.8133		0.8805	
Rounded	0.78		0.84		0.81		0.88	
Coef_G absolute	0.7032		0.7580		0.7716		0.8319	
Rounded	0.70		0.76		0.77		0.83	
Relative error variance	0.0079		0.0051		0.0063		0.0037	
Relative SEM	0.0889		0.0714		0.0792		0.0609	
Absolute error variance	0.0115		0.0087		0.0081		0.0055	
Absolute SEM	0.1074		0.0934		0.0899		0.0743	

SEMs for fixed differentiation facets will inform us about the variability in the observed data, it cannot do so for a future application of the design. Should unbiased prospective estimates be needed, then the G study analysis should be re-run after changing the status of Methods to infinite random.

G-facets analysis table

In some sense, optimizing measurement precision by increasing the numbers of levels of facets is a convenient strategy, because it can result in an increase in measurement reliability in the absence of detailed knowledge about exactly what impacts negatively on this. However, as noted in Chapter 2, in some applications much of the measurement error can be attributable to the atypical behavior of one or more specific levels of an instrumentation facet: for example, an item answered correctly relatively more frequently by weaker students than by able ones, a class in which certain learning activities are more popular than in the majority of others, and so on. Such interaction effects can at times indicate flawed items and at others item bias (however this is defined). A G-facets analysis can show precisely the extent to which each level of an instrumentation facet affects the relative and absolute coefficients. A facet analysis is requested by checking the *G-facets analysis* box at the bottom of the Workscreen (see Figure 3.1), at which EduG displays a list of those instrumentation facets for which a facet analysis can be carried out, for the user to select those of interest. Note that differentiation facets are, by definition, not amenable to a G-facets analysis, since they contribute only to true score variance. Nested facets will also not be included in the facets list, since "level 1," "level 2," and so on of a nested facet differ in meaning from one level of the nesting facet to another (think of the classes within teaching methods A or B).

The G-facets report gives the values of the relative and absolute G coefficients that are obtained when each of the levels of the selected instrumentation facet is in turn excluded from the analysis. To "improve" measurement precision, one can exclude particular items from a repeat analysis, or from a rescoring, to eliminate interaction effects. If a distinct feature can be identified that clearly characterizes the minority of items that behave differently from the majority, then this might help refine the definition of the facet universe that should be sampled in the assessments, at which point the questionable items can be validly excluded. A new random sampling of items from within this newly defined domain would then be advisable, rather than proceeding with the possibly unexplained set of "best behaved" items. The results appear in the form shown in Table 3.6. Items 3 and 6 should be particularly noted, since they tend to

Table 3.6 G-Facets Analysis Results for the Comparison of Teaching Methods

Facet	Level	Coef_G relative	Coef_G absolute
I	1	0.753	0.682
	2	0.771	0.721
	3	0.825	0.767
	4	0.786	0.734
	5	0.774	0.725
	6	0.839	0.769
	7	0.712	0.634
	8	0.784	0.728
	9	0.729	0.685
	10	0.775	0.746

reduce the reliability of the whole set of 10 items. Excluding them from the analysis would raise the value of Coef_G. Without item 3, the reliability of the remaining set of nine items would be 0.83 for relative measurement and 0.77 for absolute measurement. Without item 6, the corresponding coefficients would be 0.84 and 0.77, respectively.

But we repeat the word of warning offered in Chapter 2. A measurement improvement strategy of this type should only be adopted with great caution. This is because it can in practice reduce the validity of the measurement tool or procedure, by changing the nature of the universe of generalization. The strategy should be used only when theoretical and/or technical reasons justify it: for example, if it is clear that a test item is poor for some observable reason, or that the class in question had been subject to a very different curriculum experience than the others. Otherwise increased measurement precision will be at the cost of reduced measurement validity. In this case, as in others, the phenomenon might simply have resulted as an artifact of item sampling. This is what we would conclude in the absence of any obvious reason for discarding the two "aberrant" items. On this basis, the two items would not be rejected, and reliability would instead be increased in future applications by using more randomly sampled items in the questionnaire.

Before closing this discussion of G-facets analysis, the question of possible bias in the values presented in a G-facets table deserves consideration. Should we equally avoid here the effect of Whimbey's correction when applied to a finite random differentiation facet, as was done for the SEM? Should we re-compute the G-facets table? We would say not. This is because all the entries in Table 3.6, and others like it, are values of Coef_G. Their reduced size is as legitimate as the reduced value of ω^2 in comparison to $E\hat{\rho}^2$ when computed on the same data. Moreover, optimization procedures should be conducted using the same estimation design as the one used to estimate the original G coefficients. It is these reliability coefficients that the G study intends to improve. If we changed the design of the D study, the reference point would be lacking for the optimization design.

chapter four

Applications to the behavioral and social sciences

In this chapter, we describe and discuss a number of G theory applications set in the fields of education and psychology. The examples, which are simplified adaptations of real-life applications originally described in Bain and Pini (1996), serve to identify the different types of measurement challenges that can arise in practice and how the information provided by EduG can be used to address these challenges. Each example follows the same structure. After briefly describing the application, we identify the purpose of the G study. We then outline the three designs that guide the data analysis (observation, estimation, and measurement), before presenting the G study results. A general interpretation of the results is followed by a discussion of one or two issues that the particular application highlights. The chapter also covers some technical and methodological issues that need to be understood by practitioners if G theory is to be properly applied.

EduG files relating to the various example applications (files containing the data, their description, and analysis instructions) are provided with the software. Readers can thus familiarize themselves with the technical aspects of the procedure through direct data analysis, should they wish. For self-instructional purposes, it would be useful to try to design each study—establishing the observation, estimation, and measurement designs—after having read the overview of each application and the G study aims but *before* moving on to the design details, results, and interpretation. Here is a list of the example applications that we will be describing along with the name of the associated EduG file: to find the files follow *File/Open/ForPractice* in the EduG command window.

Ex 1	Measuring clinical depression in psychiatric patients	*File*: 08DepressionScaleE
Ex 2	Evaluating the selection process for business trainees	*File*: 09InterviewE

Ex 3   Determining the relative difficulty of   *File*: 12MathematicsE
       fraction items
Ex 4   Identifying mastery in handling fractions   *File*: 12MathematicsE
Ex 5   Evaluating a new method of   *File*: 11TeachingWritingE
       writing instruction
Ex 6   Assessing the development of a   *File*: 10AttitudeChangeE
       cooperative attitude

Example 1: Measuring clinical depression in psychiatric patients

The context

In a study involving a group of psychiatric patients the requirement was to measure the state of clinical depression of each of 20 patients using a psychological depression scale. An appropriate diagnostic test comprising 12 supposedly homogeneous items (high internal consistency) was administered by the same clinician to the 20 patients. The following are examples of the items used:

- Sundays, I stay at home, unless someone comes and picks me up.
- I wait for hours before making a difficult phone call.

   Subjects responded by indicating the frequency of each described behavior using a 5-point Likert scale: *nearly always, frequently, sometimes, rarely,* and *practically never*. Total test scores and mean item scores were then computed (i.e., the subjects' global results were located on a 12–60 total score scale or a 1–5 mean score scale). The higher the score, the more serious the state of clinical depression. For illustration, the items and summary scores of two of the patients are shown in Table 4.1.

Purpose of the G study

The principal purpose of the G study was to establish whether the 12-item questionnaire could reliably differentiate among the patients. This is a

Table 4.1 Items and Summary Scores for Patients 1 and 14

Items	1	2	3	4	5	6	7	8	9	10	11	12	Total score	Mean score
Patient 1	1	2	1	2	2	2	1	2	2	1	3	2	21	1.75
Patient 14	4	4	4	4	3	5	5	3	4	5	3	4	48	4.00

relative measurement. But we can also explore the reliability of an absolute measurement in a G study. Here, then, the purpose of the G study was to check the reliability of these two kinds of measures.

Setting up the G study

Two facets are involved in this application: namely, Patients (P) and Items (I). Since all 12 items were responded to by every one of the 20 patients, both Patients and Items are crossed facets. The observation design is therefore PI. Moreover, since one can imagine the existence of an infinite population of patients and of an infinite (or extremely large) population of items, any of which might have been included in the study, the two facets are each considered as infinite random. In other words, we assume that the 20 patients and 12 items actually used in the study were randomly selected from within their corresponding infinite universes (estimation design). Finally, since the purpose of the exercise is to establish the state of clinical depression of patients, P is the differentiation facet and I is the instrumentation facet, giving the measurement design P/I. The design information offered to EduG in the work screen is reproduced in Table 4.2.

Variance partition and attribution diagrams

Figure 4.1 is the variance partition diagram for this situation (note that, as mentioned in Chapter 2, the highest order interaction variance in this and other designs is actually confounded with residual variance—that is, with variance arising from unidentified sources of error—but we follow Brennan [2001, p. 24] in omitting the "e" that Cronbach et al. [1972] used to represent this residual error variance, in expressions such as PI,e).

Figure 4.2 illustrates the G study measurement design of interest, namely, P/I (vertical hatching indicates the single differentiation facet P). If relative measurement is the concern, then there is only one contributor to error variance, and this is the interaction effect PI (confounded with all unidentified variance sources). If absolute measurement is the concern, then there is an additional contribution to error variance, and this is the effect I.

Table 4.2 Declaring Facets for the Clinical Depression Study

Facet	Label	Levels	Universe
Patients	P	20	INF
Items	I	12	INF

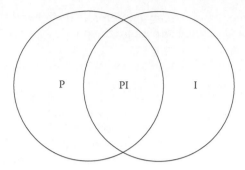

Figure 4.1 Variance partition diagram for the estimation design PI, where P and I represent Patients and Items, respectively, and both facets are random.

Results

ANOVA

The ANOVA results that relate to the variance partition diagram in Figure 4.1 are presented in Table 4.3. Since this is a random model design, the "Random" column in Table 4.3 provides the estimated variance components that we need for the G study analysis (in this application the columns "Mixed" and "Corrected" are redundant, and simply repeat the same values). The "%" column shows each component as a percentage of the sum of all components.

G study

The results of the G study analysis for the measurement design illustrated in Figure 4.2 are shown in Table 4.4. This table layout is standard for all G

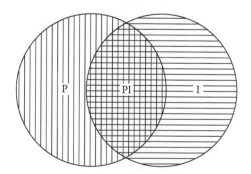

Figure 4.2 Variance attribution diagram for absolute measurement for the design P/I, where P and I represent Patients and Items, respectively, and both facets are random.

Table 4.3 ANOVA Table for the Clinical Depression Study

Source	SS	df	MS	Components			%
				Random	Mixed	Corrected	
P	131.167	19	6.904	0.543	0.543	0.543	46.0
I	58.333	11	5.303	0.246	0.246	0.246	20.8
PI	81.833	209	0.392	0.392	0.392	0.392	33.2
Total	271.333	239					100%

study analyses. The differentiation facet P is clearly shown on the left of the table, along with its variance component. The sources of generalization variances, PI for both relative and absolute measurements and I for absolute measurement, are shown on the right, with their adjusted variance components (component divided by facet sample size). The "%" columns show the relative contributions of each source of generalization variance to the total error variance for relative and absolute measurements.

It is interesting to note that while the variance component for P is identical in Tables 4.3 and 4.4, the figures shown alongside each generalization facet are not. This is because in Table 4.4 the estimated variance component shown in Table 4.3 for each generalization facet has been weighted, by dividing it by the facet's sample size, that is, by the number of levels representing that facet in the G study, to provide the contribution to one or both

Table 4.4 G Study Table for the Clinical Depression Study
(Measurement Design P/I)[a]

Source of variance	Differentiation variance	Source of variance	Relative error variance	% Relative	Absolute error variance	% Absolute
P	0.543		—		—	
	—	I	—		0.020	38.5
	—	PI	0.033	100	0.033	61.5
Sum of variances	0.543		0.033	100%	0.053	100%
Standard deviation	0.737		Relative SE: 0.181		Absolute SE: 0.230	
Coef_G relative	0.94					
Coef_G absolute	0.91					

[a] Grand mean for levels used: 3.167; variance error of the mean for levels used: 0.049; standard error of the grand mean: 0.222.

types of total error variance. Thus the variance components in Table 4.3 for I (Items, 0.246) and for PI (the Pupils-by-Items interaction variance, 0.392) are each divided by the number of items administered to each patient (12), giving 0.020 and 0.033, respectively, as shown (rounded to three decimal places) in the "absolute error variance" column in Table 4.4.

As Table 4.4 shows, the G coefficients are comfortably high in this application (greater than 0.90), inviting the conclusion that the 12-item diagnostic test has satisfactorily measured the patients' clinical states of depression: relative measurement if the interest is in comparing one individual with another and absolute measurement if the aim is rather to establish the exact location of each individual on the depression scale. In this example, individual scores are located on a 1–5 scale. Given the length of the scale, we can see that the standard errors (or "likely" errors) associated with the individual results are relatively small: the square roots of the total relative or absolute error variances (0.032 and 0.053) give standard deviations of 0.18 and 0.23, respectively, for relative and absolute measurements. These two standard errors are certainly much lower than the estimated standard deviation (0.74) for the distribution of "true" scores (square root of the estimated total differentiation variance, 0.543).

Illustration of model symmetry

We can use this example to provide a concrete illustration of the symmetry of the generalizability model. We have mentioned earlier that in a situation such as this the practitioner could pose a number of quite different questions. Here, for example, one could have been interested in item differentiation rather than patient differentiation. The question of interest would then be to check whether items could be clearly distinguished in terms of the frequency at which they were accepted by the patients as describing themselves. An item with a relatively high mean score (around 4, for example) would indicate an aspect that was more often a source of problems than one represented by an item with a much lower mean score (around 2, for instance).

An appropriate G study analysis would have the same observation and estimation designs, since the two facets are unchanged in nature and relationship; the ANOVA results therefore remain the same (Table 4.3). But the measurement design would be different—this would become I/P rather than P/I, as before, with the facets of differentiation and instrumentation having changed roles (Figure 4.3), with the patients now becoming the "instrument" used to evaluate the "sensitivity" of the items.

The analysis results for this alternative G study are given in Table 4.5, which shows that the G coefficients for item measurement both have

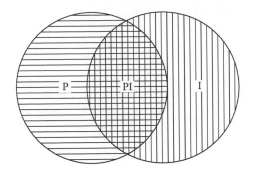

Figure 4.3 Variance attribution diagram for absolute measurement for the design I/P, where P and I represent Patients and Items, respectively, and both facets are random.

acceptably high values, even if lower than for the measurement of patients. If item evaluation had indeed been the focus of the study, one could conclude that, for the population of psychiatric patients that the 20 assessed patients represented, the different items were indeed able to indicate states of depression occurring with varying degrees of frequency (item means varying from 2.5 to 3.8 on the 1–5 scale).

The G coefficients for the two symmetric applications described here happen to be very similar, but this need not always be the case. It is quite possible, and even quite common, for the coefficients to be high for one application (differentiating individuals or differentiating items) and much lower for the other.[i]

Table 4.5 G Study Table for the Clinical Depression Study
(Measurement Design I/P)

Source of variance	Differentiation variance	Source of variance	Relative error variance	% Relative	Absolute error variance	% Absolute
	—	P	—		0.027	58.1
I	0.246		—			
	—	PI	0.020	100	0.020	41.9
Sum of variances	0.246		0.020	100%	0.047	100%
Standard deviation	0.496		Relative SE: 0.140		Absolute SE: 0.216	
Coef_G relative	0.93					
Coef_G absolute	0.84					

Example 2: Evaluating the selection process for business trainees

The context

A business school given responsibility for selecting and training business managers for a client company adopted the following practice. Once a year, service managers in the company identified a number of junior staff that they considered would be capable of assuming particular managerial responsibilities, after suitable training. The nominated candidates (120 in the study described here) were individually interviewed by two psychologists from the business school, who independently rated them for each of four attributes: personality, motivation, intelligence, and social adaptability. The eight ratings made of each candidate (two raters on four attributes) were based on a common scale running from 1 (*poor*) to 4 (*excellent*). Table 4.6 shows the ratings given to one candidate by the two psychologists.

A mean score was then calculated for each candidate, by averaging over the eight attribute ratings (for instance, the candidate shown in Table 4.6 has a mean score of 2.63). The next step was to order the candidates from best to worst in terms of their mean scores, before accepting into the training scheme as many of the highest scoring candidates as there were training places available.

Purpose of the G study

The G study was intended to establish the extent to which the selection procedure could reliably rank order training scheme candidates in terms of average attribute ratings. The reader is encouraged to think about whether the appropriate scale of measurement in this application is relative or absolute. The question of reliability coefficients for relative and absolute measures will be discussed later in this section.

Setting up the G study

There are three facets involved here: Candidates (C), Interviewers (I), and Attributes (A). Since every candidate was rated on all four attributes by

Table 4.6 Attribute Ratings for One of the Candidates

	Personality	Motivation	Intelligence	Adaptability
Interviewer 1	3	2	3	2
Interviewer 2	2	2	3	4

Table 4.7 Declaring Facets for the Business Trainee
Selection Study

Facet	Label	Levels	Universe
Candidates	C	120	INF
Interviewers	I	2	INF
Attributes	A	4	4

both interviewers working independently, all three facets are crossed (observation design CIA).

As to the estimation design, the two facets C and I can be considered infinite random. This is because both facets have a theoretically unlimited number of levels (in the case of candidates, the universe includes all company employees who were, or would be, eligible for training in years other than that considered here), and it can be assumed therefore that the levels "observed" in this study—the particular candidates evaluated and the two psychologists who carried out the evaluations—represent random samples from within these infinite universes. In contrast, the Attributes facet universe has just four levels, since all the attributes considered relevant for this type of candidate selection feature in the study. For this reason the Attributes facet is a fixed facet, with no level sampling having been employed. As the G study is intended to establish the reliability of candidate ordering, Candidates is the single differentiation facet, with Interviewers and Attributes being the generalization facets. The measurement design is therefore C/IA. This entire set of design information is entered in the EduG command window as shown in Table 4.7.

Variance partition and attribution diagrams

The variance partition/attribution diagram, which as usual combines the observation and estimation designs, is shown in Figure 4.4 (the fixed status of the Attributes facet is indicated by a dotted circle). The corresponding ANOVA results are given in Table 4.8.

Figure 4.5 represents the measurement design for the G study in question: as usual, vertical hatching highlights the differentiation facet, which on this occasion is Candidates. In the case of relative measurement, the type of measurement at issue here, the interaction effect CI will contribute to the error variance, but interaction effects CA and CIA will not, because A is a fixed facet. This point concerning interactions with fixed facets will be discussed later. For absolute measurement the effects I and CI will both contribute to error variance.

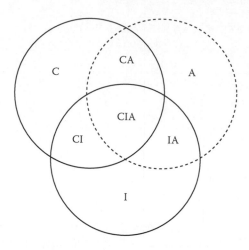

Figure 4.4 Variance partition diagram for the estimation design CIA, where C, I, and A represent Candidates, Interviewers, and Attributes, respectively, and A is a fixed facet.

Results

ANOVA

A point to note about the data in Table 4.8 is the fact that the three columns of estimated variance components show different values for different effects. Differences in values between the "Random" column and the "Mixed" column are due to the fact that we are here dealing with a mixed model design, in which the facet Attributes is fixed. The random model

Table 4.8 ANOVA Table for the Business Trainee Selection Study

Source	SS	df	MS	Components			%
				Random	Mixed	Corrected	
C	109.8240	119	0.9229	0.0222	0.0317	0.0317	8.4
I	7.5260	1	7.5260	0.0120	0.0143	0.0143	3.8
A	3.4531	3	1.1510	−0.0007	−0.0007	−0.0006	0.0
CI	79.5990	119	0.6689	0.1248	0.1672	0.1672	44.5
CA	87.9219	357	0.2463	0.0382	0.0382	0.0287	7.6
IA	3.7615	3	1.2538	0.0090	0.0090	0.0068	1.8
CIA	60.6135	357	0.1698	0.1698	0.1698	0.1273	33.9
Total	352.6990	959					100%

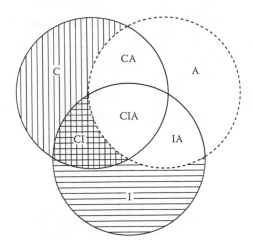

Figure 4.5 Variance attribution diagram for the measurement design C/IA, where C, I, and A represent Candidates, Interviewers, and Attributes, respectively, and A is a fixed facet.

values would apply if all three facets were random facets (as in the previous example). Differences in component values between the "Corrected" column and the "Mixed" column result from application of Whimbey's correction (see Chapter 2) to those components that are associated with effects involving the fixed facet Attributes.

Note in Table 4.8 the small negative value of the Attributes variance component estimate. Negative estimates can be surprising when first encountered, since variances can in theory have only positive or zero values. Rest assured! Variances are indeed always positive or null. Where negative variance estimates emerge they are typically close to zero, as here. They occasionally occur because we are dealing with sample-based component estimates, calculated on the basis of observed mean squares, and when a true component value is close to zero its estimate can take a small negative (or positive) value. When the estimate is negative, EduG follows convention by automatically replacing it with zero (in parenthesis), not brackets. In subsequent phases of analysis the component is treated as null, and so has no impact on the precision of the measurements.

On a final note, we mention here that should a negative variance component emerge for a differentiation facet, rather than for an instrumentation facet, then the corresponding differentiation variance is to be considered as null. This means that if there is one single differentiation facet in a study design and if its variance component is null, then the G study analysis will be meaningless, because the coefficients of generalizability for relative and absolute measurements will be zero.

Table 4.9 G Study Table for the Business Trainee Selection Study
(Measurement Design C/IA)

Source of variance	Differentiation variance	Source of variance	Relative error variance	% Relative	Absolute error variance	% Absolute
C	0.0317		—		—	
	—	I	—		0.0071	7.9
	—	A	—		(0.0000)	0.0
	—	CI	0.0836	100.0	0.0836	92.1
	—	CA	(0.0000)	0.0	(0.0000)	0.0
	—	IA			(0.0000)	0.0
	—	CIA	(0.0000)	0.0	(0.0000)	0.0
Sum of variances	0.0317		0.0836	100%	0.0908	100%
Standard deviation	0.1782		Relative SE: 0.2892		Absolute SE: 0.3013	
Coef_G relative	0.28					
Coef_G absolute	0.26					

G study

The G study results are presented in Table 4.9, which shows that the G coefficients, of relative and of absolute measurement, are both below 0.3, falling well short of the conventionally accepted value of 0.80. This means that the selection procedure was not able to produce a reliable ordering of candidates on the basis of their mean attribute ratings. Only around 28% of the observed variation in mean ratings is attributable to candidates' "true scores" (as indicated in the relative G coefficient, whose value is 0.28), leaving around 72% attributable to random unidentified effects that have seriously reduced the precision of individual results.

Coefficients of relative or of absolute measurement?

In this application the aim of the exercise was to produce an ordering of candidates in which the only thing that counted was the relative position of each candidate in the final ordered list, based on average attribute ratings. Since candidate selection was achieved by identifying those candidates with the highest average ratings, in numbers dictated by the number of training places available each year, what needed to be assured was that each candidate was reliably placed within the ordered list. Nonsystematic interviewer biases that might affect some candidates more than others (shown as the interaction effect CI) would contribute to error variance. But any rating bias that applied equally to all candidates (effect I) would have no impact on the validity of the final ordering, and would therefore not contribute to error variance. The appropriate coefficient for indicating the reliability of candidate positioning within the final list is the G coefficient for relative measurement (0.28). As a measure of the power of the design, this value indicates that the measurement procedure used, with just two interviewers independently rating candidates, has not produced valid candidate ranking outcomes.

The situation would have been different had candidate selection depended on whether or not mean ratings matched or exceeded a predetermined criterion value: for example, a mean rating of 3.0 or better. In this case the precision of the mean ratings themselves would have been the focus of interest, and the appropriate coefficient to calculate would be the coefficient of absolute measurement (0.26); here, any rating bias could risk resulting in a different selection decision for some or all candidates. The effect I of evaluator biases would contribute to the error of measurement in this case, in addition to the interaction between candidates and interviewers (CI).

Why such low reliability coefficients?

A careful look at the values shown in Table 4.9 suggests two explanations for the low reliability coefficients. Firstly, the standard deviation associated with the differentiation facet is very low, at 0.18 on a measurement scale of 1–4. This means that rating differences between candidates were small, either because most of the candidates were indeed very similar in their characteristics or because the procedure used was unable to detect the differences that did exist (it is relevant to note that 70% of the individual rating results—the candidates' mean ratings—fell in the range 3–4, with just 3% lower than 2.5, the mid-point of the rating scale). The second factor to impact negatively on the value of the coefficients is the size of the interaction effect between candidates and interviewers (CI), an effect accounting for 92% of the total absolute error variance. This interaction is the principal source of imprecision in the measurements, and is evident from the fact that the positions of individuals within the ordered list could change considerably depending on the rater. Thus, even though the rating averages of the two psychologists (2.96 and 3.14, respectively) would suggest that their *overall* standards of judgment were closely similar, their judgments of individual candidates were sometimes quite different. In fact, differences in the psychologists' ratings of individual candidates over the four attributes varied between −1.25 and +1.25. One of the psychologists was more severe than the other when rating certain candidates and more lenient when rating others.

Null error variances: Why?

On a final note, this example offers the opportunity to comment on a small but important technical point concerning null error variances. As Table 4.9 shows, the error variances associated with several of the effects, namely, A, CA, IA, and CIA, are shown as zero values in parenthesis. The fixed nature of the Attributes facet holds the explanation for this phenomenon. As we have already indicated in Chapter 2, a fixed facet that forms part of the face of instrumentation is not affected by random sampling fluctuations. Consequently, a fixed instrumentation facet cannot impact measurement precision, either directly or through interaction effects. The variance components associated with corresponding effects (non-null, as seen in Table 4.8, the ANOVA table) are thus ignored in later phases of the analysis (one speaks here of "passive" variance, symbolized in Figure 4.5 by the blank areas inside the A circle). However, and this distinction is important, the interaction between a fixed *differentiation* facet and one or more *random* instrumentation facets contributes both to absolute and to relative measurement errors (for an example, see the illustrative application discussed in Chapters 1 and 2—the comparison of mathematics teaching methods).

SEM and significance testing

The low reliability of the interviewer results is so disappointing that one might question whether the interviewers really could differentiate the candidates at all, and maybe this hypothesis should be tested. The design associated with this example application is exceptionally simple, having two crossed infinite random facets (facet A, being fixed, has no effect). It is the only type of G study considered in this chapter in which an F test might readily be applied, as mentioned in Appendix D. That is why we discuss it further here, even though we believe that the whole point of G coefficients is to indicate the power of a design for different types of measurements, rather than to test the statistical significance of particular factor effects. Conventionally we might test the statistical significance of the difference between Candidates by using an F test, based on the ANOVA results, with the mean square for Candidates (0.9229, from Table 4.8) in the numerator of the F-ratio and the mean square for the interaction Interviewers by Candidates (0.6689) in the denominator. A glance back at Figure 4.5 confirms that such a test will be valid, since the C circle contains the CI interaction and no other (as Attributes has been declared fixed, this facet's interactions with C and I do not contribute any random variance and can therefore be neglected). The F-ratio is 1.38, which just reaches statistical significance at the 5% level with 119 degrees of freedom both for the numerator and the denominator.

The G study analysis, on the other hand, offers a different kind of information, in the form of the SEM for candidate scores, with which confidence intervals can be produced if we assume an underlying normal distribution. From Table 4.9 we see that the "relative" SEM is 0.2892. The 95% confidence interval around a candidate mean score is then ±0.567. If we turn our attention now to differences between candidate scores, we need to find the SEM for a generic difference score. On the assumption of independence of measures (realized with these residual scores), we can calculate the variance of a difference score as equal to twice the "relative" variance shown in Table 4.9, that is, twice 0.0836, or 0.1672. The square root of this value, that is, 0.4089, is the standard error of the difference that we are looking for. The 95% confidence interval around a difference score is then 1.96 times 0.4089, that is, 0.80. Such a difference would occur with a probability of 0.05 or lower, which means that should any such large differences emerge in the interview selection process, we would be fairly secure in accepting them as "real."[ii]

Example 3: Determining the relative difficulty of fraction items

The context

A research study carried out in a junior high school (seventh to ninth grades) aimed to determine the relative difficulty of 10 fraction items (Bain & Pini, 1996, pp. 80–83). The items were specifically chosen to provide a representative sample of the fraction handling skills that students should have acquired by this point in their mathematics education. It was particularly interesting to compare the purely mathematical skill required to carry out operations such as these:

- 1/6 + 1/4
- 8/9 − 5/6
- 2/3 × 3/10
- 1/7 ÷ 1/3

with the ability to solve word problems like this:

- You have a 1 m long canvas. You use one quarter of it for a cushion. What length of material do you have left?

To control for possible order of presentation effects, the researchers created two versions of each test (versions A and B), the order of presentation of the 10 items being randomly determined in each case. The study involved 120 randomly selected eighth grade students: 60 from the "academic" stream (students destined for higher academic education) and 60 from the "nonacademic" stream (students generally destined for vocational education). Within each group half the students were given version A of the test and half version B. The items were scored 1 for a correct answer and 0 otherwise. Thus, an item correctly solved by 90 of the 120 students would have a facility value of 0.75 (i.e., a mean score of 0.75 on the 0–1 scale, meaning that 75% of the students would have answered it correctly).

Purpose of the G study

The principal aim of the G study was to check whether the measurement procedure used would provide a precise estimate of each item's facility. But there were three additional questions of interest:

- Was the overall test result (the mean score over all 10 items) more or less the same for the two student groups (the two streams) or not?
- Was there evidence of an item presentation order effect on the degree of difficulty of the test as a whole (i.e., version A vs. version B)?
- Was the relative position of items in the difficulty hierarchy the same or different from one student group to the other?

Setting up the G study

The observation design involves four facets: Groups (G) formed by the two streams, test Versions (V), Students (S), and Items (I). The facet Items is crossed with the other three facets: the same items appeared in both versions of the test and were administered to all the pupils in the two groups. Similarly, the facet Groups is crossed with the facet Versions: both versions of the test were administered to both groups of students. On the other hand, the facet Students is nested both in the facet Groups and in the facet Versions, since each student belonged to one group only and was administered one version only of the test. Being simultaneously nested within two crossed facets, we say that the facet Students is nested in the intersection of Groups and Versions (S:GV). The observation design is therefore (S:GV)I.

As far as the estimation design is concerned, the facet Groups is a fixed facet since the only two possible levels (academic and nonacademic) are both included in the study. But as in the previous examples, the facets Items and Students are infinite random (randomly sampled from universes that are in principle infinite). Although theoretically finite (since the number of permutations of a series of 10 items is equal to 10!), the universe of the facet Versions is nevertheless considered infinite, because the number of possible orderings is particularly high: of the order of 3.6 million. An alternative choice could be that the facet Version is fixed.

Table 4.10 Declaring Facets for the Item Difficulty Study

Facet	Label	Levels	Universe
Groups	G	2	2
Versions	V	2	INF
Students in Groups and Versions	S:GV	30	INF
Items	I	10	INF

Since the aim of the study is to verify that the measurement procedure is able to differentiate among items of this type in terms of their relative difficulty, or item facility, the measurement design is I/GVS. The design details are offered to EduG as shown in Table 4.10.

Note that when defining facet nesting, relationships are indicated using a colon (as above, S:GV indicates that Subjects are nested within Groups and Versions). In the absence of colons EduG treats all other facets as crossed facets. When the members of a nested hierarchy are crossed with an "external" facet, this latter facet appears in as many interaction effects as there are facets within the nested hierarchy it is crossed with. In this particular example, for instance, the facet I is crossed with the nested hierarchy S:GV, and the relevant interaction effects are therefore IG, IV, IGV, and I(S:GV). (Since G, V and S:GV were declared before I in the EduG Workscreen [Table 4.10], EduG uses the equivalent notation GI, GV, GVI, and SI:GV in results tables.)

Variance partition and attribution diagrams

The variance partition diagram for this situation is shown in Figure 4.6, while the measurement design is shown in Figure 4.7. Since G (Groups, or streams) is a fixed facet, the Groups main effect cannot contribute to relative or to absolute error variance, and neither can any interactions involving this facet directly. Thus, as Figure 4.7 illustrates, if the aim of the exercise had been to establish how well the fraction items could be located relative to one another on a scale of difficulty, then the only two potential sources of error variance would be the interactions between Items and Versions, and between Items and Students, that is, IV and SI:GV. For absolute error variance we would also have Versions and S:GV (Students within Groups by Versions).

Results

ANOVA

The ANOVA results are given in Table 4.11. By far the greatest contribution to the total score variance, at just over 60%, comes from Student

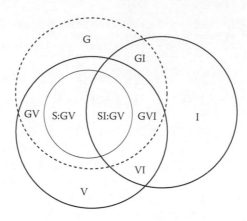

Figure 4.6 Variance partition diagram for the estimation design (S:GV)I, where S, G, V, and I represent Students, Groups, Versions, and Items, respectively, and G is a fixed facet.

interaction with Items—denoted in Table 4.11 by SI:GV. Remember, though, that this is the component that is confounded with any as yet unidentified sources of variance and with random error. Interitem variation (I) is the next largest contributor, followed by Students within Groups and Versions (S:GV). Note that since V has a small negative component estimate this potential source of error variance, in both relative and absolute measurement, becomes irrelevant, and a null value is carried forward into the G study analysis (since G is a fixed facet, GVI would not anyway have been a source of error variance in either case).

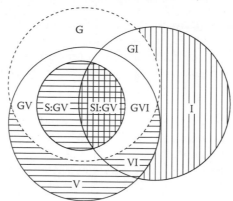

Figure 4.7 Variance attribution diagram for relative measurement for the design I/GVS, where S, G, V, and I represent Students, Groups, Versions, and Items, respectively, and G is a fixed facet.

Table 4.11 ANOVA Table for the Item Difficulty Study

Source	SS	df	MS	Components Random	Mixed	Corrected	%
G	24.653	1	24.653	0.040	0.040	0.020	7.8
V	0.003	1	0.003	−0.001	−0.001	−0.001	0.0
S:GV	53.873	116	0.464	0.031	0.031	0.031	12.0
I	53.897	9	5.989	0.048	0.048	0.048	18.9
GV	0.750	1	0.750	0.001	0.001	0.001	0.2
GI	1.130	9	0.126	0.000	0.000	0.000	0.1
VI	1.680	9	0.187	0.001	0.001	0.001	0.2
SI:GV	162.927	1044	0.156	0.156	0.156	0.156	60.9
GVI	0.967	9	0.107	−0.002	−0.002	−0.001	0.0
Total	299.880	1199					100%

G study

Let us first recall that the objective of this study was to verify if the measurement procedure used provided a precise measurement of item facility for the different fraction items. The interpretation of the results will therefore focus on the absolute coefficient of reliability. As Table 4.12 shows, this coefficient is very close to 1, indicating that the procedure was able very satisfactorily to estimate item facility. So it is possible not only to position the items reliably in a difficulty hierarchy (from easiest to most difficult—relative measurement), but also to establish with precision their actual facility values. One could then say that for the student population under consideration, such and such an item has, for example, a score of 0.25 (25% student success rate), of 0.50, of 0.82, and so on: absolute measurement.

In a similar way, we note that (on a 0–1 scale) the SEM is around 0.04 (4%), which confirms the high degree of reliability (of precision) achieved for the measurement in this particular situation. On a more technical note, we also draw attention to the fact that the two coefficients are very close to one another. This is explained by the fact that the only variance source that in principle contributes to the absolute coefficient but not to the relative coefficient, that is, Students within Groups and Versions (S:GV), is relatively unimportant, representing just about one-seventh (14%) of the absolute error variance and only 0.5% of the total variance (differentiation plus generalization variance). G is a fixed facet, and so contributes to neither coefficient, while V has a null component of variance.

Applying generalizability theory using EduG

Table 4.12 G Study Table for the Fraction Items Study
(Measurement Design I/GVS)

Source of variance	Differentiation variance	Source of variance	Relative error variance	% Relative	Absolute error variance	% Absolute
	—	G	—		(0.0000)	0.0
	—	V	—		(0.0000)	0.0
	—	S:GV	—		0.0003	14.2
I	0.0484		—		—	
	—	GV	—		(0.0000)	0.0
	—	GI	(0.0000)	0.0	(0.0000)	0.0
	—	VI	0.0003	16.4	0.0003	14.1
	—	SI:GV	0.0013	83.6	0.0013	71.8
	—	GVI	(0.000)	0.0	(0.0000)	0.0
Sum of variances	0.0484		0.0016	100%	0.0018	100%
Standard deviation	0.2199		Relative SE: 0.0394		Absolute SE: 0.0426	
Coef_G relative	0.97					
Coef_G absolute	0.96					

Additional questions

In addition to the general conclusion just offered, here are the answers that further analysis provides to the three additional questions of interest.

1. *Was the overall test result (the mean score over all 10 items) more or less the same for the two student groups or not?*
   If we look at Table 4.11, the ANOVA table, we see that the variance associated with the facet Groups is in no way negligible, having a mean square of 24.65 and making a substantial contribution to the total expected variance (of around 8% if the facet is considered random and 14% if it is considered fixed). This suggests the existence of a relatively large difference between the means of the two groups. Indeed, the test means were, respectively, 0.37 and 0.65 on the 0–1 scale. The global level of difficulty of the test clearly differed from one group to the other: on average, 37% of the items were correctly answered by the nonacademic group and 65% by the academic group. If we conduct a G study, with the measurement design G/SVI (Figure 4.8 shows the new variance attribution diagram), we find that the two means are differentiated with a reliability of 0.95 for relative

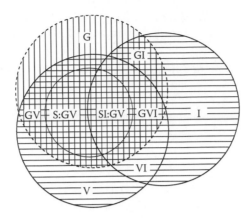

Figure 4.8 Variance attribution diagram for absolute measurement for the design G/SVI, where S, G, V, and I represent Students, Groups, Versions, and Items, respectively, and G is a fixed facet.

measures. But the absolute coefficient is equal to 0.77 only. The average score for each group is thus not known with high precision.

The SEM for absolute measurement of these two means is 0.077 (Table 4.13). But this SEM was computed for a fixed differentiation facet, and is thus inadequate for estimating the expected standard error, as explained in Chapters 2 and 3. If we recompute the SEM, by reclassifying Groups as an infinite random facet and rerunning the G study, we actually find little difference in this particular case. The absolute SEM becomes 0.079 in place of 0.077. This is partly due to the fact that several sources of variance contribute to the absolute variance expression in this design, rendering it, and its square root, more robust to a change in the status of the differentiation facet than the relative SEM. But it is also due to the fact that the modified sampling status of Groups results in some variance contributions being transferred from the relative to the systematic parts of the error variance. Absolute error, being the sum of the two, is not affected by the different variance attribution. As a matter of fact, it is difficult to predict what will be the effect of fixing a facet previously considered random, or changing the definitions the other way round. It is the components of the facets crossed with the one that is changed that will be increased or decreased, and some estimates will not differ, due to compensating changes.

2. *Was there evidence of an item presentation order effect on the degree of difficulty of the test as a whole (version A vs. version B)?*
   Similar reasoning seems to suggest that the effect of the facet Versions was insignificant. In fact, the variance generated by this facet is very

Table 4.13 G Study Table for the Comparison of Groups in the Fraction Items Study (Measurement Design G/SVI)

Source of variance	Differentiation variance	Source of variance	Relative error variance	% Relative	Absolute error variance	% Absolute
G	0.0199		—		—	
	—	V	—		(0.0000)	0.0
	—	S:GV	0.0005	48.1	0.0005	8.7
	—	I	—		0.0048	81.6
	—	GV	0.0003	26.1	0.0003	4.7
	—	GI	0.0000	1.4	0.0000	0.3
	—	VI	—		0.0000	0.4
	—	SI:GV	0.0003	24.4	0.0003	4.4
	—	GVI	(0.0000)	0.0	(0.0000)	0.0
Sum of variances	0.0199		0.0011	100%	0.0059	100%
Standard deviation	0.1411		Relative SE: 0.0327		Absolute SE: 0.0770	
Coef_G relative	0.95					
Coef_G absolute	0.77					

small, indicating that success on the test (or, equivalently, the level of facility/difficulty of the whole set of items) was not appreciably affected by the order in which the test items were presented (mean square 0.003, error variance 0). The mean scores for the two test versions confirm this conclusion, with values, respectively, of 0.512 (on average 51.2% of the items correctly answered) and 0.508 (50.8%) for versions A and B.

3. *Was the relative position of items in the difficulty hierarchy the same or different from one student group to the other?*
Phrased in slightly different terms, this question requires study of the interaction between Groups and Items (GI). In the example discussed here this effect is particularly weak (mean square 0.750, corrected component of variance 0.00015) and, moreover, its contribution to the error variance is practically zero. We conclude, then, that the item difficulty hierarchy hardly varies from one group to the other, as one could equally confirm by calculating a correlation coefficient (e.g., Spearman's ρ[iii]) between item positions when ordered from easiest to most difficult for the two groups (ρ = 0.95).

Example 4: Identifying mastery in handling fractions

The context

Suppose that in the previous study, with its four facets G, V, S:GV, and I, the purpose of the student testing was not so much to establish the difficulty levels of individual fraction items, but was rather to identify students as successfully demonstrating, or not, that they had acquired the relevant fraction handling skills at the end of a mastery learning program. The aim of the exercise was to identify those students (irrespective of their stream) whose mastery was judged insufficient, so that they might be given remedial activities, and those whose mastery was clearly demonstrated, so that they might be given more challenging work. The mastery criterion was decided to be a test score of at least 65% on whichever of the two test versions a student attempted.

Purpose of the G study

In this case, the G study was to find out whether the test was able precisely to locate students with respect to the threshold score (criterion) that separated satisfactory results from those judged unsatisfactory. The criterion being independent of the distribution of the students' scores, we will see below that absolute (rather than relative) measurement is at stake here.

Setting up the G study

For this analysis, we note right away that since the testing situation is one and the same, the observation and estimation designs are unchanged (see Figure 4.6). There are four facets, with unchanged numbers of levels and unchanged crossing/nesting relationships: 10 fraction Items (I) in two test Versions (V), two Groups (G)—academic and nonacademic, and 30 Students within each Group by Version (S:GV). Groups is a fixed facet, but Students, Items, and Versions are considered infinite random. The only change is to the measurement design. To define this, remember that when a nested facet appears on the face of differentiation (Students in this case) the respective nesting facets must necessarily appear there as well. Since the facet Students is nested in the facets Groups and Versions, the measurement design is GVS/I (or, equivalently, SGV/I, SVG/I, VSG/I, VGS/I, or GSV/I). Figure 4.9 illustrates the new measurement design.

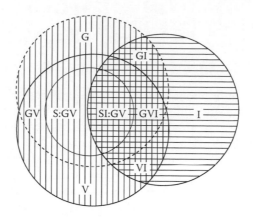

Figure 4.9 Variance attribution diagram for absolute measurement for the design GVS/I, where S, G, V, and I represent Students, Groups, Versions, and Items, respectively, and G is a fixed facet.

Results

For the reasons just mentioned, the ANOVA results are the same as before (Table 4.11), but the G study results change (Table 4.14).

Since the intention was to locate students as precisely as possible with respect to the cut-score (65% of the maximum possible test score), the absolute coefficient is a more appropriate choice of reliability index than the relative coefficient. But Table 4.14 shows that the value of the absolute coefficient is not high, indicating that the measurement was not very satisfactory, mainly because of the high interaction between Students and Items (SI:GV—76% of the error variance). In consequence, there is a relatively high degree of uncertainty associated with the placing of students (or at least some of them) on the scale. What we really want, however, is to obtain a dependable measure of the distance (positive or negative) of each student from the cut-score. Fortunately, there is another approach to the problem of judging the adequacy of the procedure in this respect. This specific reliability measure is discussed below, under "criterion-referenced coefficient."

Standard error, margin of error, and confidence interval

We have observed earlier, in Chapters 2 and 3, that the SEM is a useful, sometimes essential, element to consider when interpreting G study results. But the use of EduG is frequently oriented toward design optimization. In these cases, it is the expected value of the error variance in a future application of the design that interests us. To estimate it, the

Table 4.14 G Study Table for the Fractions Mastery Study
(Measurement Design GVS/I, with G a Fixed Facet)

Source of variance	Differentiation variance	Source of variance	Relative error variance	% Relative	Absolute error variance	% Absolute
G	0.0199		—		—	
V	(0.0000)		—		—	
S:GV	0.0308		—		—	
	—	I	—		0.0048	23.6
GV	0.0006		—		—	
	—	GI	0.0000	0.1	0.0000	0.1
	—	VI	0.0001	0.3	0.0001	0.2
	—	SI:GV	0.0156	99.6	0.0156	76.1
	—	GVI	(0.0000)	0.0	(0.0000)	0.0
Sum of variances	0.0513		0.0157	100%	0.0205	100%
Standard deviation	0.2265		Relative SE: 0.1252		Absolute SE: 0.1432	
Coef_G relative	0.77					
Coef_G absolute	0.71					

differentiation facet must be declared of infinite size. The result in the present case is 0.1256 for the relative SEM and 0.1435 for the absolute SEM. Differences between these values and the values given earlier (Table 4.14) are negligible, but they are nevertheless important to be aware of.

The absolute SEM is equivalent to one-seventh of the scale length (0–1). This confirms the imprecise, and rather unstable, character of the test scores, a feature that becomes even clearer if we use the standard error to construct a margin of error around the cut-score in question. First, remember that the success criterion adopted by the researchers was a test score of at least 65%. On the 0–1 scale used by EduG this translates to a score of 0.65. The 95% confidence interval around this criterion score, created by adding and subtracting from it roughly twice (strictly 1.96 times) the SEM (i.e., 0.1435 × 1.96 = 0.28), is 0.37–0.93. In principle, students scoring below the minimum value of the confidence interval, getting, say, at most four items right, would definitely be considered to have demonstrated an inadequate level of mastery, and could be identified for remedial activities. Those scoring above the highest value, that is, above 0.93, which in practice means getting at least nine items correct, could be given more challenging activities. Those within the confidence interval would not be able to be categorized reliably one way or the other, and would either need to take a longer test or be judged on the basis of other evidence. In practice, however, no one would use a 10-item test with these properties to make judgments of any kind about individual students, and certainly not judgments about their future instruction. A longer test should have been used if this was the aim of the assessment exercise.

Absolute coefficient or criterion-referenced coefficient?

For the reasons mentioned earlier, given the task at hand, the aim of the study was less about getting precise measurements of the mastery levels actually reached by the students than it was about accurately locating each of them on one side or other of the cut-score (0.65). Given this, it might be that the absolute coefficient is too "severe," because it takes no account of the location of the score distribution (or of its mean) on the measurement scale. If the overall mean test score is far from the cut-score (a value of 0.2 or 0.3, for example), then most of the students could be assumed to be reliably evaluated (in terms of their position with respect to the cut-score), even if their "true" scores were not very precisely estimated. On the other hand, if the mean test score was close to the cut-score (a mean test score of 0.6, for instance) it would need much more precise measurement in order to make reliable decisions about whether individual "true" scores were above or below the cut-score. The criterion-referenced coefficient, $\Phi(\lambda)$, is entirely appropriate for addressing this issue (see Chapter 2), because it not only takes into consideration differences between individual scores and the overall mean score in the definition of differentiation variance, but also the distance between the overall mean score and the cut-score, λ.[iv] In this example, the absolute coefficient has a value of 0.71 and the criterion coefficient a value of 0.76. EduG provides the criterion-referenced coefficient once a cut-score has been identified. In our example, with an overall test mean score of 0.51 and a cut-score of 0.65, the value of $\Phi(0.65)$ is 0.76 (this estimate of reliability was based on Groups being defined as fixed, with two levels; if groups had been declared as infinite random, then the criterion-referenced coefficient would have been equal to 0.80).

"Composite" differentiation variance

A word or two about those elements that contribute to differentiation variance will be in order at this point. We observe in the second column of Table 4.14 that this variance (0.0513) is obtained by summing the variances of four different sources, respectively the facets and facet interactions G (Groups), V (test Versions), GV (Groups by Versions), and S:GV (Students nested in the Group by Version interactions). This situation can be analyzed as follows. The presence on the differentiation face of the source S:GV is not surprising, since the between-student variance is directly linked to the aim of the study: to obtain reliable measurement of individual student results. The effect of facet G is equally relevant, because it represents the global difference in success between the two groups. From this perspective, the deviations observed between students for the two groups considered together have two components: interstudent differences within each group on the one hand and intergroup differences on the other (differences in success within each group are increased by the difference that exists between the groups). Finally, for similar reasons, the effects V (Versions) and GV (Groups by Versions) also appear in the expression for differentiation variance, because they amplify even more the variability in the individual student results (as it happens, given their null or very low values, these two components can be ignored in this particular case).

Observation design reduction

Suppose now that a separate G study is needed for each of the two student groups. The results already obtained from a global G study provide a good idea of what the new analyses might show. As we have pointed out in the previous section, the differentiation variance is in reality a combination of variance contributions from four different sources, among which the facet Groups is the major contributor. Now, if we take each group separately, then the facet Groups (for which just one level will be retained in the analysis) and all its interactions will have null variances. Consequently, with the Versions effect still assumed to be negligible, the only contribution to the differentiation variance will come from variance between students within the group concerned. In addition, since there is no reason to suspect that the variance associated with the facet Items and its interactions will be markedly changed (the interaction GI in the previous analysis being very weak), we can assume that the G coefficients will tend to be lower, since the differentiation variance will be proportionally weaker compared with the error variance and with the total variance.

To verify the validity of this reasoning, a separate G study analysis was carried out for each student group. The data set used earlier was used again for these new analyses, but with a reduced observation design: EduG worked successively with the data from one and then the other of the two levels of the facet Groups (see Chapter 3 for details of how to request an observation design reduction). The G study results for Group 1 are shown in Table 4.15. As we see, the G coefficients for Group 1 are 0.65 (relative) and 0.59 (absolute). The same analysis undertaken for Group 2 produces closely similar values of 0.67 (relative) and 0.61 (absolute) (Table 4.16).

Table 4.15 G Study Table for Group 1 of the Fractions Mastery Evaluation (Measurement Design GVS/I)

Source of variance	Differentiation variance	Source of variance	Relative error variance	% Relative	Absolute error variance	% Absolute
G	(0.0000)		—		—	
V	(0.0000)		—		—	
S:GV	0.0291		—		—	
—		I	—		0.0051	24.8
GV	(0.0000)		—		—	
—		GI	—		—	
—		VI	0.0003	1.9	0.0003	1.4
—		SI:GV	0.0150	98.1	0.0150	73.8
—		GVI	—		—	
Sum of variances	0.0291		0.0153	100%	0.0204	100%
Standard deviation	0.1705		Relative SE: 0.1238		Absolute SE: 0.1427	
Coef_G relative	0.65					
Coef_G absolute	0.59					

As could have been predicted, the G coefficients for each group separately are lower than they were for the analysis that involved both groups. A review of the analysis results shows that this drop is principally attributable to a lowering of the differentiation variance, in which the components G and GV are no longer present. The total error variances (relative and absolute) are hardly changed: the interactions GP and GVI, which have disappeared as a consequence of the design reduction, were already practically nonexistent, and the variances associated with other sources of error have remained more or less the same. On a purely intuitive basis, the conclusions drawn from the analysis are easy to understand. In effect, it is harder to distinguish among the students in the more homogeneous within-group samples than it is within the student sample as a whole.

Full and reduced designs: what are the educational implications?

Apart from the differences that we have discussed concerning the coefficients obtained from analyses based on full and reduced observation designs, it is important to understand that these two approaches relate to different situations and to different pedagogical perspectives. In the first case, the aim was to identify, among all the students in the age

Table 4.16 G Study Table for Group 2 of the Fractions Mastery Evaluation (Measurement Design GVS/I)

Source of variance	Differentiation variance	Source of variance	Relative error variance	% Relative	Absolute error variance	% Absolute
G	(0.0000)		—		—	
V	0.0001		—		—	
S:GV	0.0326		—		—	
—		I	—		0.0046	22.3
GV	(0.0000)		—		—	
—		GI	—		—	
—		VI	(0.0000)	0.0	(0.0000)	0.0
—		SI:GV	0.0162	100.0	0.0162	77.7
—		GVI	—		—	
Sum of variances	0.0328		0.0162	100%	0.0208	100%
Standard deviation	0.1810		Relative SE: 0.1272		Absolute SE: 0.1443	
Coef_G relative	0.67					
Coef_G absolute	0.61					

group concerned, and independently of whether they were in an academic stream or not, those students who would benefit from educational support. This would be useful only if the learning objectives were the same for both groups, and the same support mechanism was considered equally relevant. In the second case, evaluation (and possibly also support) issues would be considered separately for each group and would have different implications for educational follow-up. One could guess, then, that if in a much more homogeneous group the test is incapable of clearly differentiating levels of mastery (optimization attempts having failed), then measures other than a single identification of supposedly weak students would have to be adopted, for example a formative evaluation and more individualized teaching within normal lessons.

Example 5: Evaluating a new method of writing instruction

The context

A new method of teaching the skills involved in producing position papers was tried with sixth grade students in half of the 26 schools in a particular school district (Bain & Pini, 1996, pp. 75–79). At the end of the academic year those institutions involved in the initiative wanted to evaluate the effectiveness of the method, in terms of the quality of students' persuasive writing. Two of the 13 schools that had participated in the experiment were selected at random for inclusion in the evaluation study, as were two of the 13 institutions that had continued to use the so-called traditional teaching method (the control group). Three classes were randomly selected in each of these four schools. Every student in every class was asked to produce a position paper defending the following proposition, which was related to the minimum age for getting a driving license:

> In rural areas, students living far from their schools and facing long hours of travel each day should get special permission to shorten their trip by driving part of the way.

The number of students in each class varied from 15 to 20. To satisfy the EduG constraint of equal numbers of levels of a nested variable within the different levels of a nesting variable, 15 students were randomly selected from within the larger classes so that all classes were represented in the evaluation study by 15 students (an alternative would have been to retain all the data, and to use some other statistical software to compute sums of squares for the unbalanced data set, these sums of squares then being fed into EduG—see Chapter 3 for details). The students' writing was separately rated for each of the following three qualities (marks 1–6 each):

- Relevance of the arguments offered
- Linking and organization of arguments
- Spelling and syntax.

The resulting student profiles each comprised three separate skill ratings.

Purpose of the G study

The aim of the G study was to find out whether the measurement procedure used permitted a reliable comparison of the average writing skills of pupils taught by one or the other method. Relative measurement is clearly relevant here, but absolute measurement would also be of interest.

Setting up the G study

Five facets feature in this application: Methods (M), Institutions (I), Classes (C), Students (S), and Skill Domains (D). Of these five, the first four

Table 4.17 Declaring Facets for the Persuasive Writing Study

Facet	Label	Levels	Universe
Methods of teaching	M	2	2
Institutions in Methods	I:M	2	13
Classes:Institutions:Methods	C:I:M	3	INF
Students:Classes:Institutions:Methods	S:C:I:M	15	INF
Domains	D	3	3

constitute a single nesting hierarchy: Students nested in Classes, Classes in turn nested within educational Institutions (schools), and Institutions nested within Methods (individual schools being entirely allocated to one or other method). The facet Domains, however, is crossed with every other facet, since all students, in every class in every institution, whatever method they were taught by, were rated for the same three writing qualities. The observation design is then (S:C:I:M)D.

As far as the estimation design is concerned, we started by considering the facets Methods and Domains as fixed. The Institutions facet is a finite random facet, since two schools were randomly selected from a finite population of 13 institutions, for both the new method and the old. As to Classes and Students, these will be considered as usual as infinite random facets (while in reality the numbers of students and classes available for selection at the time of the study were both finite, we consider these simply to be members of an infinite population of similar students and classes: past, present, and future). The aim was to see how well, if at all, the two teaching methods could be distinguished in terms of students' writing outcomes. The measurement design is therefore M/ICSD. The design details are offered to EduG in the usual way (Table 4.17).

Variance partition and attribution diagrams

The variance partition diagram is shown in Figure 4.10, while the particular measurement design analyzed is illustrated in Figure 4.11.

For relative measurement, Figure 4.11 confirms that when Methods is the differentiation facet, all of its nested facets contribute to relative measurement error (Institutions, Classes, and Students). Interactions with the facet Domains, on the other hand, do not contribute to the relative error variance. This is because Domains is a fixed facet, so that neither it nor any of its interactions with other facets can be a source of random fluctuation (see the blank areas in Figure 4.11).

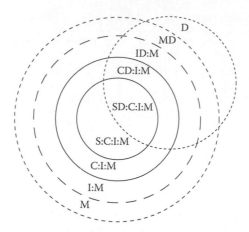

Figure 4.10 Variance partition diagram for the estimation design (S:C:I:M)D, where S, C, I, M, and D represent Students, Classes, Institutions, teaching Methods, and Domains, respectively, and M and D are fixed facets.

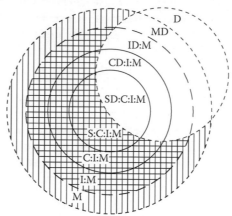

Figure 4.11 Variance attribution diagram for the measurement design M/ICSD, where M, I, C, S, and D represent teaching Methods, Institutions, Classes, Students, and Domains, respectively, and M is a fixed facet.

Results

ANOVA

The ANOVA results are given in Table 4.18, in which we once again see the emergence of negative component estimates, this time for Institutions within Methods (I:M) and for Classes interacting with skill Domains (CD:I:M). The most interesting feature in Table 4.18, however, must be the very high contribution of the between-student effect (S:C:I:M) to total

Table 4.18 ANOVA Table for the Persuasive Writing Study

Source	SS	df	MS	Components Random	Mixed	Corrected	%
M	55.424	1	55.424	0.179	0.183	0.091	5.0
I:M	10.433	2	5.216	−0.033	−0.033	−0.030	0.0
C:I:M	77.659	8	9.707	0.155	0.150	0.150	8.1
S:C:I:M	495.333	168	2.948	0.787	0.982	0.982	53.3
D	119.700	2	59.850	0.320	0.330	0.220	11.9
MD	4.225	2	2.113	0.018	0.018	0.006	0.3
ID:M	1.688	4	0.422	0.001	0.001	0.001	0.1
CD:I:M	5.451	16	0.340	−0.016	−0.016	−0.010	0.0
SD:C:I:M	196.933	336	0.586	0.586	0.586	0.390	21.2
Total	966.850	539					100%

variance, at 53%. The contribution of the interaction between students and domains (SD:C:I:M, albeit confounded with all unidentified sources of variance) is also noteworthy, at just over 20%.

G study

The G study analysis results are presented in Table 4.19. The first feature to note in this G study is that the relative error variance results from nested facets, and not as in earlier examples from interactions between the differentiation facet(s) and crossed generalization facets. The second point worthy of note in Table 4.19 concerns the values of the G coefficients. In all the examples presented earlier, the coefficient for relative measurement has always been higher than the coefficient for absolute measurement. Here, though, we note right away that the relative and the absolute reliability coefficients have the same value, 0.72. This situation is neither fortuitous nor unusual. It has happened in this application for reasons explained below.

Contribution of a nested facet to relative error: An anomaly?

In a G study, the relative error variance is attributable to interaction effects between the differentiation facet and one or more generalization facets, whereas main effects (in the ANOVA sense) associated with generalization facets contribute only to the absolute error variance, as do interaction effects not involving the differentiation facet. Now, in the example discussed here, where M (Methods) is the differentiation facet, we have seen that between-class variability (C:I:M) and between-student variability

Table 4.19 G Study Table for the Persuasive Writing Study
(Measurement Design M/ICSD)

Source of variance	Differentiation variance	Source of variance	Relative error variance	% Relative	Absolute error variance	% Absolute
M	0.0917	—	—		—	
	—	I:M	(0.0000)	0.0	(0.0000)	0.0
	—	C:I:M	0.0250	69.6	0.0250	69.6
	—	S:C:I:M	0.0109	30.4	0.0109	30.4
	—	D	—		(0.0000)	0.0
	—	MD	(0.0000)	0.0	(0.0000)	0.0
	—	ID:M	(0.0000)	0.0	(0.0000)	0.0
	—	CD:I:M	(0.0000)	0.0	(0.0000)	0.0
	—	SD:C:I:M	(0.0000)	0.0	(0.0000)	0.0
Sum of variances	0.0917		0.0360	100%	0.0360	100%
Standard deviation	0.3028		Relative SE: 0.1896		Absolute SE: 0.1896	
Coef_G relative		0.72				
Coef_G absolute		0.72				

(S:C:I:M) are between them responsible for the *relative* error variance. Should this surprise us?

In fact, the contributions of the facet Classes (nested in I and M) and the facet Students (nested in C, I, and M) to the relative error variance are explained by the very fact that these facets are nested within the facet Methods (between-school variance would have contributed as well, for the same reason, had its variance component estimate not been null). In nontechnical language, one can say that a generalization facet nested in a differentiation facet contributes to relative error because its levels are not the same from one level to another of the nesting facet. Because of this, the levels of the differentiation facet are not identically affected by the nested facet—for example, classes do not receive students of the same ability. This can result in changes in the relative positioning of the levels of the differentiation facet (some classes might be weaker than others). Expressed more technically, in situations like this, it is impossible to separate the interaction variance between the nested and the nesting facets, on the one hand, from the variance attributable to the nested facet, on the other.

Identical coefficients of relative and absolute measurement: Why?

If we look at the relative and absolute error variances in the results in Table 4.19, we see that both are determined in practice by the same two sources: C:I:M and S:C:I:M. This is what causes them to have the same value. But why is it so? The only effect that could have contributed to absolute error but not relative error is facet D, Domains, because it is the only facet that is crossed with the differentiation facet M. But this effect is null, since facet D is fixed (contributing only to passive variance). For the same reason, all interactions involving D (which in principle should contribute to both relative and absolute error) are also null. On this basis, we can say that the relative and the absolute coefficients are identical for two reasons: firstly, because the facet D is fixed, and, secondly, because the other facets are all nested in the differentiation facet M and thus contribute to the relative error variance, as explained in the preceding paragraph. The most important consequence of this equality in the two coefficients is that we have no way of distinguishing between the sources of error for a relative and for an absolute scale of measurement. Since the sources of random fluctuation affecting the precision of the measures are the same, comparisons will be done with the lower precision of the absolute scale.

A word of caution

Looking now at the actual value of the two G coefficients (0.72), it seems that the evaluation study was not able entirely satisfactorily to distinguish between the two teaching methods (remember that the generally accepted criterion value for a G coefficient is ≥0.80). On the other hand, the deviation from this usual criterion value is quite small, inviting reflection on the very meaning of the criterion and the use to which it is put. Here, as elsewhere in statistics, the adoption of decision criteria based on so-called critical values could lead one to believe that the set of all possible results is divided into distinct and qualitatively different classes. In reality, such criteria are at least in part arbitrarily determined on a continuous scale, where differences (especially small differences) are more of degree than substance. A difference of a few hundredths between a coefficient that is slightly higher than the criterion value and one that is slightly lower is generally not an important or meaningful difference as far as the qualities and defects of the measurement tool or procedure under evaluation are concerned. It is therefore better to avoid semi-automatic interpretations that lead to simple conclusions of the type good or bad, acceptable or unacceptable, reliable or unreliable. In most cases the conclusion should be reached on the basis of a more considered analysis of the situation.

Here in particular, we should not forget that the Coef_G that is reported in Table 4.19 is in fact an ω^2 (see Chapter 3), since the differentiation facet was declared fixed. As explained in Chapter 2, with two levels only the variance estimate is twice as large with a random sample as it is when the universe has been declared fixed. Here the variance component for Methods is 0.09 when the facet is considered fixed and 0.18 when it is considered infinite random. Coef_G is 0.72 in the first case and 0.84 in the second. Which one is correct?

In fact both values are "correct." But they do not describe the same reality. With a finite universe, the coefficient describes for a fixed facet the actual proportion of true variance in the total score variance. With an infinite universe, the coefficient *estimates* what the true variance proportion would be in a large series of experiments, should these be possible to conduct. In the finite universe case we measure the actual size of the observed effect, the intensity of the experimental factor's influence. In the case of an infinite universe, the estimate indicates the average value of the design if it could be applied to random samples of experimental and control groups, but it says nothing about the importance of the particular observed levels, because these will change each time the design is implemented. In the type of situation considered here, we suggest that ω^2 is the preferred indicator. But we are ready to admit that in other situations, where the preparation of future D studies is the focus, the ordinary intraclass correlation ω^2 could be more appropriate.

Although formally unsatisfactory, the coefficients obtained in this example indicate that the measurement strategy was not *entirely* inadequate, suggesting by the same token that a few modifications might bring reliability up to the standard normally required. This leads us into the important optimization phase.

Optimization phase

In general, we focus on those facets that contribute the most to error variance when seeking to improve an evaluation or measurement procedure. One commonly used strategy is to increase the number of sampled levels representing these facets. In this regard, EduG's optimization function provides estimates of the coefficients that would be obtained should the levels of one or more generalization facet be changed. In some cases, practical reasons may justify reducing the sample size of one facet in favor of another, while maintaining roughly the same total number of data points. Trial and error might help to find the best combination.

Here, the major sample-based fluctuations are attributable to the facets Students and Classes, as we have seen. But the number of students involved in the study is already the maximum number available (15 per class, to meet the balanced design requirement), leaving no scope for increase. On the other hand, the number of classes sampled within each school could have been increased. This would in any case have been the most appropriate strategy, since a high 70% of the error variance is in fact attributable to differences between classes within schools. One can thus estimate, for example, the likely reliability of the evaluation procedure should first four classes, and then five, rather than three, be involved in each of the two schools following each method. The output from an optimization analysis is a new report that displays the recalculated relative and absolute G coefficients that could be expected in each of the two hypothetical cases: 0.77 in the first case (four classes per school) and 0.81 in the second (five classes per school). Increasing the numbers of classes involved in each school does, therefore, result in an increase in the G coefficient. This reflects a functional relationship that is particularly easy to describe. Figure 4.12 shows the curve relating the number of classes per school to the reliability of measures (relative and absolute, since they coincide in this case), based on the Spearman–Brown formula (which estimates the reliability of a test composed of k' items on the basis of its reliability with k items, as explained in Keyterms, Chapter 2, note iv).

The interesting feature in Figure 4.12 is the slow increase in reliability for each class added into the study. Clearly, there will be a limit to the size of increase that can be achieved overall, and also a point beyond which any further small increase would not be cost-effective. In practice also, a limit is determined by the lowest number of classes available in any one of the schools, given the requirement for data balance.

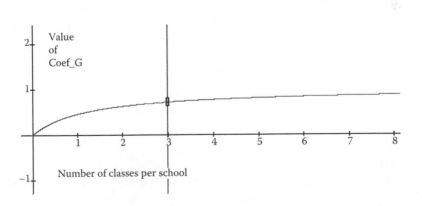

Figure 4.12 Expected effect of sampling more classes (Spearman–Brown formula).

Differentiation of domains

If the G study had been carried out to differentiate among the mastery levels reached in the three competence domains (measurement design: D/MICS), the following variance sources would have contributed to relative and to absolute error:

- *Relative error*: ID:M, CD:I:M, and SD:C:I:M.
- *Absolute error*: I:M, C:I:M, and S:C:I:M as well as ID:M, CD:I:M, and SD:C:I:M.

As shown in Table 4.20, this analysis would have resulted in two G coefficients with high values: 0.99 (relative) and 0.92 (absolute), respectively.

We note in passing that the evaluation study was better able to distinguish among the three competence domains (coefficients above 0.90) than between the two teaching methods (a coefficient equal to 0.72), the three domain mean scores (the two methods combined) being, respectively, 4.3, 3.1, and 3.6.

Table 4.20 G Study Table for the Persuasive Writing Study: Domain Comparison (Measurement Design D/MICS)

Source of variance	Differentiation variance	Source of variance	Relative error variance	% Relative	Absolute error variance	% Absolute
	—	M	—		(0.0000)	0.0
	—	I:M	—		(0.0000)	0.0
	—	C:I:M	—		0.0125	61.3
	—	S:C:I:M	—		0.0055	26.8
D	0.2201		—		—	
	—	MD	(0.0000)	0.0	(0.0000)	0.0
	—	ID:M	0.0003	10.5	0.0003	1.3
	—	CD:I:M	(0.0000)	0.0	(0.0000)	0.0
	—	SD:C:I:M	0.0022	89.5	0.0022	10.6
Sum of variances	0.2201		0.0024	100%	0.0204	100%
Standard deviation	0.4692		Relative SE: 0.0493		Absolute SE: 0.1428	
Coef_G relative	0.99					
Coef_G absolute	0.92					

Example 6: Assessing the development of a cooperative attitude

The context

A psychosocial research study was designed to explore the extent to which engagement in collaborative cognitive problem-solving activities might lead to the development of positive attitudes toward cooperation and tolerance among 14–15-year-olds (Bain & Pini, 1996, pp. 71–74). The study involved 24 randomly selected adolescents, divided at random into four groups of six. The subjects, who did not know one another before the study began, took part in five weekly hour-long sessions, all groups working on the same problem-solving tasks set by the researcher. The subjects' attitudes were measured twice, using a 20-item Likert scale—once at the start of the study and again 10 days after the last problem-solving session. Using a 1–4 response scale—*I completely agree, I tend to agree, I disagree,* and *I completely disagree*—the subjects were invited to indicate their degree of agreement or disagreement with opinions expressed in items such as the following:

- People lose time when they try to find a solution together.
- When several people attack a problem together, they run a smaller risk of remaining totally blocked.

On the basis of the responses, preexperience, and postexperience total scores were calculated for each subject, by summing the item scores (after suitably reversing the scale for some items). Total scores were on a scale of 20–80 (20 items each on a 1–4 scale); the higher the score, the more positive the individual's attitude to cooperation and tolerance. Table 4.21 presents the preexperience and postexperience item and total scores for one of the adolescents.

Purpose of the G study

The G study was intended to evaluate the extent to which the attitude questionnaire provided a reliable measurement of (assumed positive)

Table 4.21 Preexperience and Postexperience Attitude Scores for One Adolescent

Items	1	2	3	4	5	6	...	17	18	19	20	Total
Before	3	3	2	3	3	2	...	3	2	1	2	47
After	2	4	2	4	3	4	...	3	3	2	3	59

attitude development between the start and end of the problem-solving initiative. Relative measurement is clearly relevant here.

Setting up the G study

Groups (G), Subjects (S), Moments (M), and Items (I) are the four facets that feature in this study. Three of the facets, namely, Groups, Moments, and Items, are crossed facets, since every subject in every group responded to the same items on both occasions. But since each adolescent belonged to one and only one of the four groups, the facet Subjects is nested within the facet Groups (S:G). The observation design is then (S:G)MI.

As to the *sampling* of the facets, both Items and Subjects can be considered random facets, since there would in principle be an infinite (or extremely large) number of items that could have featured in the attitude questionnaire, and an infinite number of adolescents that could have participated in the research study. Groups will also be considered an infinite random facet. Moments, on the other hand, would normally be considered a fixed facet, since the only assessment moments of interest (and the only ones that could reasonably be treated within the framework of this research) are the two that actually feature in the study: the beginning and end of the problem-solving experience. Such a choice, however, has important consequences for SEM estimation, as we discuss below. For the estimation design, then, we have three infinite random facets (Subjects, Groups, and Items) and one fixed facet (Moments).

Since the aim is to check the ability of the questionnaire to reliably detect any attitude change between the beginning and the end of the research study, the measurement design is M/GSI. Table 4.22 shows the way the study is set up within EduG.

Variance partition and attribution diagrams

Figure 4.13 illustrates the variance partition for this particular combination of observation and estimation designs, while Figure 4.14 illustrates the measurement design for this G study application.

Table 4.22 Declaring Facets for the Attitude Change Study

Facet	Label	Levels	Universe
Groups	G	4	INF
Subjects in Groups	S:G	6	INF
Moments	M	2	2
Items	I	20	INF

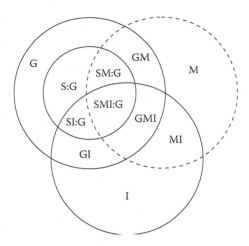

Figure 4.13 Variance partition diagram for the estimation design (S:G)MI, where S, G, M, and I represent Subjects, Groups, Moments, and Items, respectively, and M is a fixed facet.

Since S, G, and I are random facets, all interaction effects between these facets and the facet M will contribute to relative error variance, and thus will affect the size of the relative G coefficient—the coefficient of interest here. As Figure 4.14 shows, there are five such interaction effects. For absolute measurement all the main and interaction effects shown in Figure 4.14, with the obvious exception of M itself, would contribute to error variance.

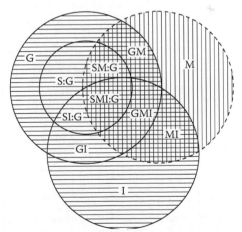

Figure 4.14 Variance attribution diagram for absolute measurement for the design M/GSI, where M, G, S, and I represent Moments, Groups, Subjects, and Items, respectively, and M is a fixed facet.

Results

ANOVA

The ANOVA table is shown as Table 4.23. As in Example 2, note once again the impact of adjusting random model component estimates to fit the mixed model that we have here, and of applying Whimbey's correction. Note also the negative corrected variance component estimates for the effects GI and MI. Another feature that is interesting to note in the data presented in Table 4.23 is the fact that just three of the potential sources of variance account for over 80% of the total variance (i.e., of the sum of corrected variance components). These are S:G, or between-student variance within groups, SI:G, student interaction with items, and SMI:G, which, remember, subsumes not only the interaction effect between Students and Items and Moments, but also all unidentified sources of variance.

G study

At the start of the study the average total attitude score for the whole group of 24 adolescents was 52.5, rising to 60.4 at the end. Over the course of the research study there was therefore an average increase of 7.9 points on the 20–80 point scale, or 0.39 points on the 1–4 scale that EduG used. But how reliably have these average scores been measured? Table 4.24 has the answer.

G coefficients

As Table 4.24 shows, just four of the five potential sources of variance that could contribute to the total error variance for relative measurement

Table 4.23 ANOVA Table for the Cooperative Attitude Change Study

Source	SS	df	MS	Components Random	Mixed	Corrected	%
G	37.895	3	12.632	−0.006	0.004	0.004	0.4
S:G	233.971	20	11.699	0.229	0.275	0.275	27.9
M	38.001	1	38.001	0.070	0.070	0.035	3.5
I	70.228	19	3.696	0.066	0.063	0.063	6.4
GM	14.286	3	4.762	0.020	0.020	0.010	1.0
GI	38.668	57	0.678	−0.019	−0.003	−0.003	0.0
SM:G	43.637	20	2.182	0.091	0.091	0.045	4.6
SI:G	269.279	380	0.709	0.171	0.354	0.354	36.0
Ml	7.686	19	0.405	−0.006	−0.006	−0.003	0.0
GMI	31.943	57	0.560	0.032	0.032	0.016	1.6
SMI:G	138.946	380	0.366	0.366	0.366	0.183	18.6
Total	924.541	959					100%

Table 4.24 G Study Table for the Attitude Change Study
(Measurement Design M/GSI)

Source of variance	Differentiation variance	Source of variance	Relative error variance	% Relative	Absolute error variance	% Absolute
	—	G	—		0.0010	4.7
	—	S:G	—		0.0114	53.8
M	0.0348		—		—	
	—	I	—		0.0031	14.8
	—	GM	0.0025	50.1	0.0025	11.7
	—	GI	—		(0.0000)	0.0
	—	SM:G	0.0019	38.1	0.0019	8.9
	—	SI:G	—		0.0007	3.5
	—	MI	(0.0000)	0.0	(0.0000)	0.0
	—	GMI	0.0002	4.1	0.0002	0.1
	—	SMI:G	0.0004	7.7	0.0004	1.8
Sum of variances	0.0348		0.0050	100%	0.0213	100%
Standard deviation	0.1865		Relative SE: 0.0704		Absolute SE: 0.1459	
Coef_G relative		0.88				
Coef_G absolute		0.62				

(0.0050) do so, with the two greatest contributors—the interaction between Groups and Moments (GM) and the interaction between Students (within groups) and Moments (SM:G)—between them accounting for over 80% of the total relative error variance. For absolute measurement, on the other hand, we have a greater number of contributors to total error variance, at eight. Among these, by far the most important is between-student variation within groups (S:G), with a contribution of just over half of the total error variance. As to the G coefficients, we see from Table 4.24 that the coefficient of relative measurement is reasonably high, at 0.88. This confirms that the questionnaire has produced a reliable estimation of the difference in attitude levels recorded for the whole group of adolescents before and after their problem-solving experience. On the other hand, the coefficient of absolute measurement is relatively low, at 0.62. This indicates that the questionnaire did not produce a particularly reliable measure of *absolute* attitude levels on each separate occasion. This is in great part explained by the high intersubject variability within the four groups, noted above, which has specifically affected the coefficient of absolute measurement but which has no effect on the coefficient of relative measurement.

In this particular study, the apparent weakness of the assessment tool for absolute measurement is not really an issue, because the research set out to estimate attitude change (i.e., the difference between average attitude scores on the two occasions) and not actual values on each occasion. As mentioned earlier, it is thus the reliability of relative measurement that is important and not the reliability of absolute measurement; and the reliability of relative measurement was very satisfactory.

Confidence intervals

As we have commented earlier, a G coefficient provides a global index of measurement reliability, that is not always easy to interpret in simple concrete terms. Another, more intuitively accessible, way of indicating measurement precision is to establish confidence intervals around the measurement of interest, as we have already noted during discussion of the previous examples.

Here, however, the design is more complex, since the facet Moments has been defined as fixed. The important consequence of this choice is that the observed general mean is treated as though it were the true mean. All observed effects in the design are then measured in terms of their distance from this mean, and the variance of these effects is computed by dividing the sum of their squared deviations by N and not by $N - 1$. As $N = 2$ in this example (the two moments), the variance due to the experimental effect is 0.0050 under the fixed model, while it would be 0.0099 under the random model, in which Moments would have been declared infinite random.

Which one of the two relative SEMs is to be preferred? Under the fixed model the value of the SEM is 0.0704, while under the random model it is 0.0996. We give preference to the first of these two options, because it is the direct result of the observations. It describes the situation, it tells us the extent of "vibration" affecting the mean of each moment, and in this scientific endeavor we think that it is this that should be remembered. If we were preparing future training courses and wanted to predict the variability of the expected outcome, then the SEM of the random model would be preferable.

chapter five

Practice exercises

For those readers interested in evaluating their new understanding of G theory and its application possibilities, this chapter offers five example exercises to work through. A different measurement situation is described in each exercise, and a number of related questions posed. While some of the questions can be answered without recourse to analysis results, others will need relevant analyses to be carried out on supplied data sets using EduG. The data sets are accessible at http://www.psypress.com/applied-generalizability-theory or http://www.irdp.ch/edumetrie/data.htm. The exercises, which vary in difficulty and challenge, can be tackled through individual effort or group collaboration, in a formal seminar context or as an at-home assignment. The most difficult questions are asterisked. While the answers to all the questions are also included here, for maximum learning benefit the answers should only be consulted once the exercise concerned has been completed.

Food testing

The context

Consumer organizations are active in product testing, providing a valuable service in the interests of public health. But who checks the monitors themselves? How do we know that their methods are reliable, and their results valid? This question motivated a Swiss statistician to analyze data published in a local newspaper that compared the quality of the *fondue* being offered on the menus of five different local restaurants. The main ingredients of fondue are melted cheese—traditionally emmental and gruyère—and white wine. Other elements are added to the chef's personal taste, with the consequence that different restaurants have their own special recipes. The five fondues were selected, their preparation observed, and the resulting products tasted by five independent judges, all of whom were chefs or cheese industry experts. The judges rated each fondue on a 1–10 scale for each of five attributes: appearance, smoothness, lightness, flavor persistence, and overall taste.

The journalist's results were presented to readers in a way that suggested that the comparative study had been scientifically conducted, even though it had not (possible carry-over effects, for example, were not controlled). That notwithstanding, the question of interest here is the discriminating value of the ratings given. Were the ratings stable and concordant enough to justify generalizable conclusions? A G study analysis will provide the answer. The rating data are contained in the file *FonduesData.txt*, with 125 ratings (five judges judging five types of fondue for five different attributes) given in the following column order: fondue rated (A–E), the judge concerned (identified as 1–5), and the ratings offered by that judge to that fondue for each of the five attributes. We suggest that you resist the temptation to plough immediately into data analysis. Instead, stop for a moment, clarify your thoughts about what exactly is the question that you need to answer, and then think how best to design a G study to provide that answer.

Let us summarize the situation once more. It was assumed that readers of the local newspaper would be interested in knowing where they might find a good fondue in their local area. The journalist's motivation was to satisfy this need for information. The statistician's interest was not very different, but one step further from the restaurant table. This individual wanted to be *sure* of knowing where to get a good fondue, and so questioned the *dependability* of the journalist's ratings. For your part, you are going to repeat the statistician's analyses, to check again that trustworthy information about fondue quality has indeed been given to the newspaper's readership. The principal question is: Can any differences in fondue quality be reliably established on the basis of the available ratings? A slightly different question might be even more interesting: Can the distance of each fondue from a threshold of minimally acceptable quality be established with some acceptable degree of certainty? Finally, can the overall quality of each fondue be reliably estimated in absolute terms?

If you have fully understood the elements of G theory presented to you in the previous chapters, you will realize that the answers to these questions depend on the values of three different G coefficients: a relative coefficient, a criterion-referenced coefficient, and an absolute coefficient. When you discover the values of these three coefficients you might be entirely satisfied and rush out to order a fondue, or you might rather be curious to explore the data further to better understand the situation. If you discover that you can trust the joint opinions of the five judges, you might ask yourself how much confidence you would place in the results of any other journalist reports on similar product testing studies. If, on the other hand, you find that the combined opinion of the five judges was not reliable enough, you will be interested in using EduG's optimization facility to identify how such a rating exercise might be improved in the future.

If you are scientifically minded, you might well want to investigate the sources of rater variation further. You could be led to analyze other factors of the experimental design, such as differences between the judges or between the attributes, and more particularly the main sources of error, namely, the interactions between the three facets, judges, attributes, and fondues. Finally, you might wonder how different the results might have been, should it have been possible to relax some of the constraints on the way each facet was sampled. If predicted outcomes are sufficiently stable, this will reinforce the degree of confidence you can have in the current outcomes.

Questions

The situation

1. What are the observation and estimation designs?
   (In other words, identify the facets and their inter-relationships, and their sampling status).
2. What are all the analyzable components of variance? Show them on a variance partition diagram.

The three global indices of reliability

3. Is it possible to rank order the five fondues in terms of quality, in a reliable way?
4. Can the raters correctly judge the distance from an "acceptable" quality value?
5. Do all the judges give the same rating to each particular fondue (averaged over the five attributes)?

Possible improvements (optimization)

6. How many judges would be needed to maximize the outcome reliability in a future product testing exercise of this type?

In-depth investigations to better understand the situation

7. Explain the results you obtain, firstly by analyzing the differentiation variance.
8. Explain the results by also analyzing the error variance.
9. Did all judges really agree? Find out by carrying out a Judges G-facets analysis.

Possible modifications of the estimation design

10.* Would the design remain adequate if the G study considered the five tested fondues as a sample from an infinite universe?
11.* Would the random selection of a subset of the rated attributes markedly affect the reliability of the measurement procedure?

Conclusion

12.* What have you learned from this exercise?

Answers

1. *What are the observation and estimation designs?*
   For the observation design, each fondue is evaluated by each judge for each attribute. Consequently, the three facets F (fondues), J (judges), and A (attributes) are completely crossed. The estimation design is more difficult to define. Restaurants offering fondue are few in number in the newspaper's distribution area. Furthermore, since the journalist actually identified each relevant restaurant, it would be logical to define the facet Fondues as fixed. But if we want to provide consumer associations with information about how to improve their product testing designs in future, then we will need to consider Fondues to be an infinite random facet, treating this first sample of five fondues as a random representation of an infinite number of fondue recipes. We therefore start with Fondues considered as a fixed facet, but will later also look at it as infinite random. The judges are the instruments of observation in this example, and they should be considered as exchangeable with any other potential judges from the same population, that is, chefs or cheese industry experts. As far as fondue attributes are concerned, the five that were rated in the study were considered the five most important in this context, and so the facet Attributes is fixed. But here again there is another possibility, and we shall see later what the effect might be of treating Attributes as an infinite random facet. The basic initial choices are made explicit in Table 5.1.

   Table 5.1 Declaring Facets for the Fondues Evaluation Study

Facet	Label	Levels	Universe
Fondues	F	5	5
Judges	J	5	INF
Attributes	A	5	5

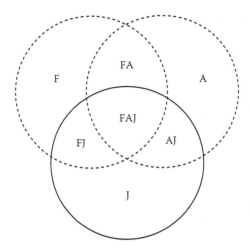

Figure 5.1 Variance partition diagram for the fondue study estimation design FJA, where F, J, and A represent Fondues, Judges, and Attributes, respectively, and F and A are fixed facets.

2. *What are all the analyzable components of variance? Show them on a variance partition diagram.*
   The three facets are crossed, and so intersect each other, leading to three 2-facet interactions and one 3-facet interaction (Figure 5.1).

3. *Is it possible to rank order the five fondues in terms of quality, in a reliable way?*
   Exceptionally, we present the full G study results table here, as Table 5.2. It confirms that the answer to this particular question is yes, given that the relative G coefficient is equal to 0.87. This proportion of true variance in the estimated total observed rating variance is quite satisfactory. The journalist has indeed published a trustworthy rank ordering of the five fondues for the newspaper's readership.

4. *Can the raters correctly judge the distance to an "acceptable" quality value?*
   For potential consumers the most important question might rather be: Can the judges reliably estimate the distance between the fondues they evaluate and a general norm of "acceptable" quality? It is easy to imagine cases where the experts might disagree about some particular attribute, but nevertheless be unanimous about which fondues were of acceptable quality and which not. Estimating the dependability of the experts' judgments about an "acceptable" fondue is equivalent to computing a criterion-referenced dependability coefficient for an examination. If we choose a λ value of six on the 1–10 rating scale we get a $\Phi(6)$ coefficient of 0.90, which is high. But a

Applying generalizability theory using EduG

Table 5.2 G Study Table for the Comparative Evaluation of Fondue Quality (Measurement Design F/JA)

Source of variance	Differentiation variance	Source of variance	Relative error variance	% Relative	Absolute error variance	% Absolute
F	0.8454		—		—	
	—	J	—		0.0450	26.7
	—	A	—		(0.0000)	0.0
	—	FJ	0.1233	100.0	0.1233	73.3
	—	FA	(0.0000)	0.0	(0.0000)	0.0
	—	JA			(0.0000)	0.0
	—	FJA	(0.0000)	0.0	(0.0000)	0.0
Sum of variances	0.8454		0.1233	100%	0.1682	100%
Standard deviation	0.9194		Relative SE: 0.3511		Absolute SE: 0.4102	
Coef_G relative		0.87				
Coef_G absolute		0.83				

λ value of seven would be too close to the general mean, leading to problems, in the sense that the value of Φ(7), at 0.78, becomes uninterpretable. It is smaller than the value of Φ itself, which is 0.83, when in theory Φ (Coef_G absolute) should be the minimum value of the two. Using the value of Φ seems to us preferable. We see that 0.83 is still comfortably high in this example.

5. *Do all the judges give the same rating to each particular fondue (averaged over the five attributes)?*
   The absolute reliability coefficient indicates this. While the relative reliability coefficient tells us if the evaluation study could rank order the fondues in a repeatable way, the absolute coefficient tells us if the various judges would agree on the precise location of each fondue on the aggregated 1–10 scale. As it happens, the value of the absolute coefficient, at 0.83, is nearly the same as the value of the relative coefficient, at 0.87. Thus the precision of absolute measurement is also sufficient. So all three global coefficients agree, confirming the dependability of the judges' evaluations.

6. *How many judges would be needed to maximize the outcome reliability in a future product testing exercise of this type?*
   Inter-rater agreement was generally high in the evaluation study. The challenge, then, is not so much to improve measurement quality

Table 5.3 Effect on Reliability of the Number of Judges

Number of judges	5	4	3	2	1
Coef_G relative	0.87	0.85	0.80	0.73	0.58
Coef_G absolute	0.83	0.80	0.75	0.67	0.50

as it is to reduce the cost of data collection by reducing the number of judges. How far could we go in this direction? EduG's optimization module gives the results, presented in Table 5.3. We see from the table that two judges would not be enough to reliably rank the fondues. It is interesting to discover that even in these exceptionally favorable conditions, relying on just one or two judges would not provide trustworthy results. Three judges, on the other hand, if drawn at random from the same population of specialists as the five that have been consulted, would in principle suffice to obtain a reliability of 0.80 for comparative judgments, but would not be sufficient for absolute judgments, given a coefficient of 0.75. Four judges would be an adequate number for this purpose. That said, the higher the reliability the better, so the original study with five judges remains superior.

7. *Explain the results you obtain, firstly by analyzing the differentiation variance.*

The quality ratings given to the five fondues are clearly different, as is shown in Figure 5.2. This means that the differentiation variance is large. The estimated true score variance component for the facet Fondues is equal to 0.8454, which represents 32% of the total variance (look at the next to last column of the ANOVA table produced by EduG—not shown here). The resulting true score standard deviation of 0.919 (square root of 0.8454) will give rise to a range of

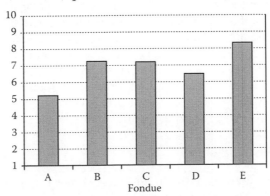

Figure 5.2 Overall rating averages for the five fondues.

scores of approximately three sigmas if the distribution is Normal, which may represent 2.5 scale points in the present case. (This is effectively the range that can be observed in Figure 5.2.) It seems to be of moderate size, if we consider the 10 points of the rating scale at the disposal of the judges, but it is nevertheless far from negligible.

8. *Explain the results by also analyzing the error variance.*
   In contrast with the differentiation variance, the relative error variance is small. This is essentially because the rated fondue attributes are considered fixed, so that between-attribute variance cannot contribute to measurement error. Only the judges can create random fluctuations in the rating data. The SEM that they produce for relative measurement is equal to 0.35. You will see from the G study table (Table 5.2) that this is caused by the interaction FJ, that is, by the fact that different judges favor different fondues. Such interactions are unavoidable, but in this case their effect is small. The interaction variance (0.1233), although equal to 73.3% of the absolute error variance, is equivalent to just 15% of the differentiation variance (0.8454).

   Perceptive readers might at this point object that the differentiation facet Fondues was defined as fixed and that, consequently, the computed SEM is smaller than it would be if the five fondues had been considered as drawn at random from an infinite population. This is true and we will explore later what difference it makes to treat Fondues as a random facet. But we were not engaged in significance testing here, nor in predicting future rating variability. We were simply comparing various sources of relative error within their common frame of reference.

   The absolute error variance is also of little importance. The SEM is not much greater for absolute measurement (0.41) than it is for relative measurement (0.35). The additional source of error in absolute measurement is the variance between the judges, some giving better ratings in general and others giving worse ratings in general. But the last column of Table 5.2 shows that this last source of random fluctuation (J) explains 27% of the absolute error variance. We can therefore conclude that the main reason for high rating reliability is that the judges were very homogeneous in their standards of judgment, following the same norms and perceiving fondue quality in the same way.

9. *Did all judges really agree? Find out by carrying out a Judges G-facets analysis.*
   A G-facets analysis of the Judges facet shows that if we temporarily exclude the ratings of Judge 1, 3, 4, or 5, the reliability of relative mean fondue rating reduces from 0.87 to values between 0.81 and

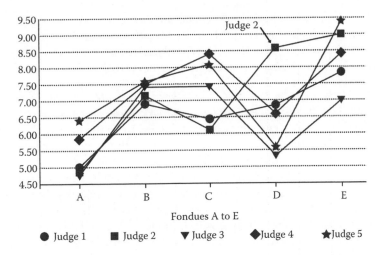

Figure 5.3 Mean attribute ratings given by each judge to each fondue.

0.85. This is expected, since eliminating even one judge would result in each fondue's mean rating being based on a smaller number of observations. But, unexpectedly, we see that if we discard the ratings of Judge 2, then the reliability of the mean rating of the other four judges increases considerably, reaching 0.92. Clearly, Judge 2's ratings were not in line with those of the other judges. This judge's ratings have introduced a relatively high degree of "noise" into the system. If we look at the results of the G-facets analysis for absolute measurement, the isolated position of Judge 2 emerges nearly as clearly. Such discordance deserves special investigation. EduG gives the mean scores for each judge for each fondue. These are presented in Figure 5.3. We see from this that Judge 2 is the only one who gave high quality ratings to fondue D. There is no obvious explanation for this particular preference, but if at the time of the study it was discovered, for example, that Judge 2 was a personal friend of the restaurateur who produced fondue D, then that would have been sufficient reason to dismiss this judge from the G study!

10.* *Would the design remain adequate if the G study considered the five tested fondues as a sample from an infinite universe?*
When we introduced this example application we mentioned that it was difficult to know how to treat the facet Fondues, in terms of its sampling status. We chose first to define it as fixed, but indicated that we would also look at what would happen if we treated it as random. The estimation design can easily be modified in the EduG

Workscreen, by changing the facet universe size from five to infinity. With Fondues random, the relative and absolute reliability coefficients become, respectively, 0.87 and 0.86, both quite satisfactory values. This was to be expected since, everything being equal, $\rho^2 \geq \omega^2$.

11.* *Would the random selection of a subset of the rated attributes markedly affect the reliability of the measurement procedure?*
The five rated attributes were considered to be the only ones of interest and consequently this facet was fixed. But this decision could be questioned. It must be possible to identify fondue attributes other than the five chosen for this study. At least we should explore the difference it might make to consider the five attributes as randomly chosen from a much larger set. Interestingly, the analysis results show that any change is limited. None of the reliability values are markedly affected, their magnitude decreasing by only 6 percentage points when the mode of sampling is changed. Of course, introducing additional random variation must increase the proportion of error variance and thus the coefficients must become smaller. With Attributes fixed, the only interaction contributing to the relative error variance is that of Judges with Fondues. With Attributes declared a random facet, the interaction Attributes–Fondues becomes a second error variance contributor, accounting for fully 30% of the new relative error variance. On the whole, however, all the sources of true and error variance remain of similar size. The optimization module shows that, with the facet Fondues random, four judges would be enough to bring the absolute coefficient up to 0.79. The gain would be the possibility of generalization.

12.* *What have we learned from this exercise?*
This relatively light-hearted example illustrates quite well the potential benefits offered by a G study for data exploration. To start with, we were able to check that the experimental design chosen by the journalist yielded trustworthy measures, because the results were coherent: the ratings of the five judges were generally in agreement. To further convince us about the value of the study, the analysis revealed a rather large differentiation variance and a very small error variance. Then a G-facets analysis identified the ratings of one of the judges as being somewhat out of line with those of the other four. Additional studies of mean ratings suggested possible reasons for the differences. Finally, the estimation design was modified twice, changing the fixed status of Fondues and of Attributes to infinite random. These changes would reflect the fact that other Fondue recipes could have been

explored in the exercise, and other quality aspects rated. In consequence, it would be possible to generalize to some extent the reliability findings of the study.

Reading ability

The context

As part of an initiative to combat youth unemployment, a small town school authority invited an educational research center to carry out a survey of the reading ability of the town's Grade 7 students. The research was to provide a population ability estimate (percentage of items correct) with a maximum standard error of 3 percentage points. The center began by designing a pilot study in order to determine an appropriate sample size for the survey proper. A total of 30 students, 15 boys and 15 girls, were drawn at random for participation in the pilot study (you will recognize that this is a simplified example for the purpose of G theory illustration—in a real-life application the number of students involved in the pilot study would count in hundreds rather than tens). The students were tested four times, with different tests assessing the same range of reading skills. The data set (with 120 data points) is organized as shown in Table 5.4. The data themselves are included in the file *ReadingScores.txt*, which contains scores only, one row per student, with no row or column headings.

The purpose of the main survey would be to estimate the overall population mean (the average reading ability of Grade 7 students), while that of the pilot study (the G study) was to determine an appropriate sample size to use in the survey itself. But this general aim leads to further questions concerning the traits that should be studied and the students who should be examined, to reduce the errors due to sampling fluctuations. The research center is also forced to consider other possible uses of the tests, in particular for student differentiation to satisfy requests

Table 5.4 Data Content of the File *ReadingScores.txt*

Subjects	Test A	Test B	Test C	Test D
Boy 01	85	52	63	89
Boy 02	81	84	80	73
⋮	⋮	⋮	⋮	⋮
Boy 15	79	62	40	91
Girl 01	80	89	63	85
⋮	⋮	⋮	⋮	⋮
Girl 15	94	63	68	81

from the teachers. It is the specialists' responsibility to make sure that such additional measures are also dependable. A G study could suggest possible changes in survey design to make such tests more reliable as individual assessments. Data analysis might also furnish other kinds of important information, such as gender differences at the item level.

Questions

The situation

1. What are the observation and estimation designs for the pilot study? (To answer, indicate how you would complete the EduG Workscreen.)
2. Which variance components will be computed? (*Hint*: Draw the variance partition diagram, identifying each area as a potential source of score variation.)

Estimating the population mean

3. What is the overall sample mean? What are the 95% confidence limits around this population estimate? (*Hint*: You can use the *Means* button on the Workscreen, although EduG also displays the overall mean immediately underneath the G study table. Here it will give the grand mean, of the four tests and of all students. The design with no differentiation facet, namely, /SGT, is especially appropriate for problems focusing on estimation of a population mean.)

4.* What would be the effect on the standard error of the grand mean if you fixed the Tests facet instead of considering it random?
5. How could you reduce the standard error of the grand mean? (*Hint*: This question can be answered by carrying out an optimization study for a measurement design focused on the grand mean and without a facet of differentiation: i.e., for the design /SGT. Try changing the number of tests.)
6. What is more efficient, increasing the number of tests or increasing the number of students?

Estimating the reliability of individual students' scores

It would be motivating for future survey students (and for their teachers) if they could be promised feedback about their own individual test performances. But such a promise could only sensibly be made if the test results were sufficiently reliable to justify it. Individual score reliability would clearly need to be investigated in the pilot study.

7. Can students be reliably differentiated on the basis of their average percentage scores across the four tests (four tests chosen randomly from within an unlimited universe)?
8. Can their individual performances be considered reliably assessed in an absolute sense?
   A less exacting objective would be to deliver a certificate of adequate reading ability to all students who achieved a minimum percentage of correct answers across the four tests.
9. What would be the cut-score that would result in a level of dependability of 0.80 for this purpose?

Complementary analyses

Gender differences

For research and policy purposes, it is interesting to know if any difference in the reading ability of girls and boys can be reliably measured.

10. What is the reliability of the test battery in terms of gender means?

Test differences

It is also important to know if the tests have distinctive features.

11. How is the third facet, Tests, measured in the pilot study?

Conclusion

12.* What have you learned from this example?

Answers

1. *What are the observation and estimation designs for the pilot study?*
   Students are nested in gender, hence S:G. The nesting facet (G) must be declared first (Table 5.5). All students take the same tests; hence

Table 5.5 Declaring Facets for the Reading Ability Survey

Facet	Label	Levels	Universe
Gender	G	2	2
Students: Gender	S:G	15	INF
Tests of reading	T	4	INF

Applying generalizability theory using EduG

the observation design is (S:G) × T. For the estimation design, Gender is a fixed facet and the facet Students is infinite random. The sampling status of the facet Tests is also purely random to make generalization possible.

2. *Which variance components will be computed?*
   The tests given to all students are the same. Hence, the facet Tests is crossed with the other two facets, Students and Gender, and therefore the three circles must intersect one another. It can be seen from Figure 5.4 that the design comprises five components of variance: G, S:G, ST:G, GT, and T.

3. *What is the overall sample mean? What are the 95% confidence limits around this population estimate?*
   You can use the *Means* button on the Workscreen here, but EduG also provides the overall mean and its standard error automatically with the G study table (unless the input data comprise sums of squares rather than raw scores). In this example you will find that the overall sample mean is 74.6. The SEM (recomputed with Gender re-classified as a random facet to provide an estimate of expected variability) is 3.355, giving a wide 95% confidence interval around the mean, of 68.0–81.2.

4.* *What would be the effect on the standard error of the grand mean if you fixed the Tests facet instead of considering it random?*
   Student performances on each of the four tests were standardized using a percentage scale. But the test means differ, and consequently the overall mean is influenced by the particular sample of tests

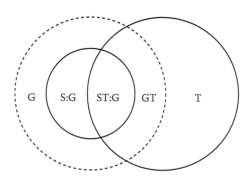

Figure 5.4 Variance partition diagram for the estimation design (S:G)T, where S, G, and T represent Students, Gender, and Tests, respectively, and G is fixed.

on which it is based. The SEM for the overall mean reflects the resulting test-related score fluctuations: the greater the difference in test means, the less precise the overall mean estimate. If the four tests are accepted as the only tests of interest, that is, if the facet Tests is considered fixed, then the variance between the test means will disappear as a contributor to error variance, and estimation precision will increase. EduG shows that the SEM reduces from 3.36 to 1.66 percentage points. But the estimate of reading ability is then limited to the four tests.

5. *How can you reduce the standard error of the grand mean?*
With this measurement design (/SGT) no reliability coefficient can be computed since no differentiation is requested. But the standard error is available, so that a confidence interval around the grand mean can be calculated. Changes in this SEM can be predicted as usual by carrying out an optimization study. With four tests, we saw that the SEM was 3.36. With five tests, the SEM would be estimated as 3.04, while with six tests it would be 2.82. The increase in estimation precision with increasing numbers of tests is quite rapid. But the practical cost of increased precision would be heavy, since four testing sessions would already be difficult to organize in the schools. There would also be a financial cost involved in constructing and printing.

6. *What is more efficient, increasing the number of tests or the number of students?*
Testing more students would be easier, but this strategy would apparently have little effect on estimation precision. With the original random sample of four tests and 30 students (15 per gender) the SEM was 3.36 percentage points. With four tests and 40 students (20 per gender) the SEM would be virtually unchanged at 3.25. With 100 students (50 per gender), the standard error would be 3.05 (the same gain as for adding a fifth test). With 200 students, the SEM would still remain near 3.00 (2.98 exactly). There would be practically no gain to expect from a further increase in the number of students tested, but the required precision of 3 percentage points could be reached with 100 boys and 100 girls.

It would, of course, be possible to modify simultaneously the number of tests given and the number of students tested. For instance, with five tests and 100 students (50 per gender), the SEM would be 2.75. But preparing a new test is costly, more so probably than testing more students. The final strategic choice would probably depend on total cost. The proposal of the researcher is to stick with the four tests and to present them to 100 students of each gender.

7. *Can students be reliably differentiated on the basis of their average percentage scores across the four tests (chosen randomly from within an unlimited universe)?*
    The answer is a resounding "No!" With a relative G coefficient of just 0.47 for student differentiation, it would be irresponsible to report individual student scores.

8. *Can their individual performances be considered reliably assessed in an absolute sense?*
    Again the answer is "Definitely not," given an absolute Coef_G of just 0.43. Some other strategy would need to be identified to motivate students and their teachers to participate in the survey.

9. *What would be the cut-score that would result in a level of dependability of 0.80 for the purpose of certificating students?*
    By trying various cut-scores, it can be shown that a criterion-referenced dependability coefficient will be equal to or greater than 0.80 for cut-scores of 88% and 61%, the overall mean being 75%. The margin of uncertainty between these two cut-scores is large, reflecting the very low precision of the individual student scores.

10. *Determine the reliability of the test battery in terms of gender means.*
    EduG shows that the boys' mean is 72.9% and the girls' mean is 76.3%. This gives a difference of 3.4 percentage points. The reliability of this difference is a problem of relative measurement. But the coefficient of relative measurement for this comparison is zero, because the variance component for Gender, the differentiation variance estimate, is negative. This suggests that the difference in gender means is small enough to be overshadowed by the variation associated with the small samples of students tested. The G study thus shows that no generalizable conclusion can be drawn about gender difference in this case.

11. *How is the third facet, Tests, measured in the pilot study?*
    With the current design, test means can be differentiated quite reliably. The relative G coefficient for comparing their mean percentage scores is 0.85. A coefficient of 0.80 could be achieved with a total sample of 28 students only (14 per gender). The standard error affecting their mean is only 2.67 percentage points.

12.* *What have you learned from this example?*
    a. This exercise illustrates the fact that reliability is not a property of tests, but is rather a property of test scores and of the use made of them. The means across the four reading tests could not

differentiate reliably among individual students or between genders, but the tests themselves could be reliably differentiated in terms of relative difficulty.

b. The overall mean is affected by the sampling of observations for each random facet in the design, and an optimizing strategy can sometimes be to increase the number of sampled levels of one facet while simultaneously decreasing the number of sampled levels of another. This gives the possibility of exploring trade-offs in representation to find an optimum balance in terms of overall cost.

c. The reason why the population mean is difficult to measure here is that the universe of tests (from which the sample of four is drawn) is very heterogeneous. More reliable results could be obtained for a narrower universe of generalization.

Mathematics

The context

Political authorities in a certain European region were under pressure to change the structure of the school system, by abolishing the selective examination that divided students into two levels at the end of sixth grade. To prepare for the expected public discussions, they asked for objective data on the region's educational achievement in the main disciplines at the end of ninth grade (but only mathematics is considered here). The report was required to provide results for the ninth grade population, and in light of these to evaluate the effect of selection at the end of sixth grade.

In response, the school authorities organized a survey of the mathematics performance of their ninth graders. To show that they were giving the same value to all students, they examined equal numbers of students in the upper and lower streams in each of the 18 existing school districts.

The students were presented with 15 mathematics items, each scored on a 0–1 continuous scale, with partial credit allowed. The data set gathered for the G study comprises results for 43 students per level (stream) per district (after random rejection of cases to introduce balance), and can be accessed by downloading the file *MathData.txt*. Data are organized in the file district by district, with 86 rows of student scores per district (43 rows of 15 scores for lower stream students and then 43 rows of 15 scores for upper stream students).

It was decided that a G study analysis should be carried out, to provide the essential information about overall student achievement in mathematics, but also to explore the effects of several identifiable factors on student achievement, specifically domain tested, district, and level/stream.

Questions

Study design

1. Give the observation and estimation designs.
   (Show how you would declare these in the EduG Workscreen.)
2. Draw the corresponding variance partition diagram.
   (Recognizing that the two streams and the 18 districts are not sampled, but fixed.)

Mathematics achievement

(In what follows you should ignore the bias resulting from the selection of an equal number of students within levels/streams and districts.)

3. Calculate the 95% confidence interval around an item mean score.
4. Can the 15 mathematics items be ordered reliably in terms of difficulty level?
5. Calculate the 95% confidence interval around the estimated population mean score.
6.* Use the optimization module to find a more adequate study design, with a more efficient balance between students and items for estimating the region's mean mathematics performance.

Performance of individual students

7. How reliably can students be differentiated, ignoring streams?
8. What essentially accounts for the differences in individual student performance?
   [*Hint*: Analyze the "true" variance with Students, Districts, and Levels (streams) comprising the differentiation face.]
9. What are the main sources of error affecting the reliability of student differentiation, and reducing the precision of individual scores?
10. Do some particular test items contribute more than others to a lack of precision in the measurement of students?
    (*Hint*: Carry out an Items G-facets analysis.)
11.* What is the reliability of student differentiation within each level?
    (*Hint*: Analyze each level separately, using an observation design reduction.)
12.* What is the criterion-referenced reliability (or dependability) coefficient for each level, using a cut-score of 0.60?
13.* Is the interpretation of the criterion-referenced coefficients the same for the two levels?
    (*Hint*: Consider the mean score for each level.)

Performance of districts

14. Is it possible for the education authority to distinguish the better-performing districts from the less good, that is, to rank districts reliably?
15. Is it possible for the education authority to measure with satisfactory precision the mean ninth grade performance of each district? What is the standard error for the district means?

Performance of levels

16.* What are the mean results for each of the two levels? How is this difference considered in G theory and in ANOVA?
17.* What distinguishes a G coefficient from an *F*-test in this case?

Conclusion

18.* What can be learned from this exercise?

Answers

1. *Give the observation and estimation designs.*
   Students are nested both in Districts and in Levels, two facets that are crossed. Hence Students are nested in the intersection of Districts and Levels, that is, S:DL. The test items were the same for all students, hence (S:DL) × I. Table 5.6 describes this structure (observation design), and also provides the estimation design. Both Districts and Levels are fixed facets, since all possible districts and both levels are included in the survey.

2. *Draw the corresponding variance partition diagram.*
   See Figure 5.5.

3. *Calculate the 95% confidence interval around an item mean score.*
   We can expect this confidence interval to be very narrow, given the large number of students taking the test. Indeed, even the absolute

Table 5.6 Declaring Facets for the Mathematics Survey

Facet	Label	Levels	Universe
Districts	D	18	18
Levels	L	2	2
Students	S:DL	43	INF
Items	I	15	INF

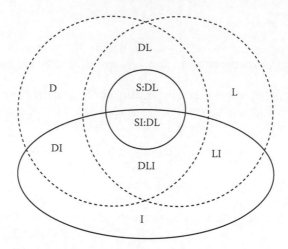

Figure 5.5 Variance partition diagram for the estimation design (S:DL)I, where S, D, L, and I represent Students, Districts, Levels, and Items, respectively, and D and L are fixed.

SEM, the most relevant here, is quite small, at 0.0073. (Note however the limited-length 0–1 mark scale.) The 95% confidence interval around an item mean score will be just 0.0145.

4. *Can the 15 mathematics items be ordered reliably in terms of difficulty level?*
   The focus of measurement here is item differentiation, and the appropriate measurement design is I/SDO. The relative G coefficient is equal to 0.998, just short of perfect. The absolute coefficient has virtually the same value, confirming that the measures of item difficulty (or of student success, as the two cannot be distinguished in G theory) are extremely precise, because of the large number of observations (students) on which they are based.

5. *Calculate the 95% confidence interval around the estimated population mean score.*
   The estimated population mean score is 0.587 on the 0–1 scale (or 58.7% on a percentage scale). The SEM for the population mean is 0.039 (or 3.9% on a percentage scale). The 95% confidence interval around the mean therefore has a lower limit given by $[0.587 - (1.96 \times 0.039)]$ and an upper limit given by $[0.587 + (1.96 \times 0.039)]$, that is, the interval is 0.51–0.66 (or 51–66% on a percentage scale). This is a very wide interval.

Questions 3, 4, and 5 seem to give contradictory information. But how can very precisely measured items be associated with such an imprecise population mean estimate? It is very important for readers to discover the reason for this apparent contradiction. The model takes into account two sources of random fluctuation: the sampling of students and the sampling of items. Question 3 shows that with 1548 students, this first source of error variance is practically controlled. But question 4 shows that the items vary quite importantly in difficulty level, with the consequence that small samples of items will be associated with large sampling variance. The next question will confirm this conclusion.

6.* *Use the optimization module to find a more adequate study design, with a more efficient balance between students and items for estimating the region's mean mathematics performance.*
Carrying out an optimization study for the measurement design (/SDOI), we find that doubling the number of items and halving the number of students, that is, changing the relative representation of items and students in the survey while preserving the same total number of observations, would reduce the absolute SEM for the overall mean to 70% of its original value (i.e., from 0.039 to 0.028). The SEM could in principle be reduced further, even to half its original value, but the number of items needed to achieve this would be impractical to administer.

7. *How reliably can students be differentiated, ignoring streams?*
If we consider all students as belonging to one common set, their results can be differentiated with a relative reliability of 0.92, and with an absolute reliability of 0.89. These values are both surprisingly high and deserve further investigation to reveal the reason for such good student differentiation. True and error variances will be studied consecutively.

8. *What essentially accounts for the differences in individual student performance?*
(As we are simply comparing components of variance and not carrying out significance tests, we can use the G parameters computed with differentiation facets declared fixed.) Since Students are nested within Districts and Levels, the differentiation variance for students has four components: D, L, DL, and S:DL. These four components sum to 0.048. The Levels component, with a value of 0.027, represents more than half of this total. The Students within Districts and Levels component (0.020) accounts for just over 40% more, while the Districts

and Districts × Levels components contribute virtually nothing. Clearly, Levels is the largest contributor to the overall between-student variance, more than individual differences within the same level. This is not surprising, given that levels (academic streams) were determined on the basis of prior student attainment.

9. *What are the main sources of error affecting the reliability of student differentiation, and reducing the precision of individual scores?*
   They are the components appearing in the column "Absolute error variance" of the G study table, namely, I, DI, LI, SI:DL, and DLI. Their sum is 0.006, of which the interaction Students × Items accounts for the largest part (70%). This shows that relative item difficulty is different from one student to another. This is often the case, but usually not to such a large extent. This source of variance is the one most responsible for the relative error. For absolute error, the component I is influential: just over a quarter of the absolute error variance is due to the large differences in overall item difficulty. Students would receive better or worse scores depending on the particular sample of items they were asked to attempt, and this obviously affects the precision of their results.

10. *Do some particular test items contribute more than others to a lack of precision in the measurement of students?*
    We discover that the differences in difficulty are important: the item means vary from 0.37 to 0.84 (Table 5.7). But no one item stands out as having a particularly large influence one way or the other on the reliability of student measurement, relative or absolute.

11.* *What is the reliability of student differentiation within each level?*
    Analyzing the response data by level, we find that for the lower level the two reliability coefficients, relative and absolute, are 0.83 and 0.77, respectively, with corresponding values of 0.82 and 0.79 for the upper level. Differences of this size between relative and absolute coefficients are often encountered in practice, but note here that the reliabilities do not differ from one level to the other. The test could reliably rank students on the attainment scale within each level. But locating each individual student precisely at a particular point on the scale was less well done. The standard error for each student mean is practically the same at the two levels, being equal to 0.07 for relative measurement and to 0.08 for absolute measurement for the lower level students, and 0.06 and 0.07, respectively, for the upper level students.

Table 5.7 Means and Contributions to Coef_G for Each Item

Item	Mean	Coef_G relative	Coef_G absolute
1	0.79	0.91	0.89
2	0.63	0.91	0.88
3	0.45	0.92	0.89
4	0.69	0.91	0.88
5	0.50	0.91	0.88
6	0.84	0.91	0.89
7	0.63	0.91	0.88
8	0.50	0.91	0.88
9	0.62	0.92	0.89
10	0.69	0.91	0.88
11	0.78	0.92	0.89
12	0.48	0.91	0.88
13	0.37	0.91	0.88
14	0.41	0.91	0.88
15	0.42	0.91	0.89

12.* *What is the criterion-referenced reliability (or dependability) coefficient for each level, using a cut-score of 0.60?*
For the lower level, $\Phi(0.60)$ has a value of 0.88. For the upper level, its value is almost the same, at 0.89.

13.* *Is the interpretation of the criterion-referenced coefficient the same for the two levels?*
Although the coefficients are the same, their meanings differ. For the lower level students, the 0.60 cut-score is toward the top end of the right-skewed score distribution, so that the test would have been very difficult for most of these students to pass. In contrast, for the upper level students, the same cut-score is situated in the lower part of the score distribution, offering no real challenge to the students concerned.

14. *Is it possible for the education authority to distinguish the better performing districts from the less good, that is, to rank districts reliably?*
The relative coefficient for District differentiation is 0.43, which is far from adequate. Thus, even though the mathematics performance of individual ninth graders could be quite reliably measured, the larger units could not be differentiated with confidence. Consequently, hasty and possibly non-valid conclusions about relative district performance should be avoided.

15. *Was it possible for the education authority to measure with satisfactory precision the mean ninth grade performance of each district? What is the standard error for the district means?*

Placing districts on a relative scale is already impossible to do reliably, as indicated by the 0.43 relative coefficient. It cannot be surprising to learn, therefore, that reliable absolute measurement is even less possible, with an absolute G coefficient of only 0.12. Yet the standard error of a district mean is only 0.043, which is quite low (thanks to the large number of students assessed, with a reasonable number of test items). Reviewing the districts' mean scores might serve to explain this apparent contradiction. The lowest district mean is 0.53 and the highest is 0.62, giving a difference of 0.09, that is, just over two standard errors. Since 18 means are contained in this quite narrow interval, it is clear that they cannot be well spread. So, for some purposes (like helping districts with substandard results) the means could be used to support decision-making, provided that district ranking is not required. (We checked that the standard errors of the district means had the same values if we declared Districts an infinite random facet.)

16.* *What are the mean results for each of the two levels? How is this difference considered in G theory and in ANOVA?*

The two levels have quite different results. The lower level mean score is 0.42, while the upper level mean is 0.75. Such a large difference might be expected, given that the level allocation of students was determined on the basis of prior attainment. The persisting difference confirms that the intended compensatory education was not successful. It is not surprising that the relative G coefficient for Level differentiation is 0.99, with a standard error equal to only 0.01. Similarly, the absolute G coefficient is 0.94 with a standard error of 0.04. With a large systematic variance and a small sampling variance, the proportion of true variance that the two coefficients express is necessarily high. Even when computed with components of variance uncorrected for facet size, the standard errors are not different.

In ANOVA, describing these data is more complex. EduG's ANOVA table gives the mean squares for all effects, in particular for the main factors L (638.05), D (0.8229), I (35.4185), and S:DL (0.3571). The design allows us to test the statistical significance of the differences between Items and between Students (within District and Level), differences that are, unsurprisingly, highly significant in this sense.

17.* *What distinguishes a G coefficient from an F-test in this case?*

Standard ANOVA practice is to compute F-tests, which, like G coefficients, are ratios of variance components, with a function of the systematic variance in the numerator and a function of the sampling variance in the denominator. How do the two techniques differ? Basically, an *F*-test computes the probability that two estimates of variance could come from the same population. A G coefficient goes one step further, by evaluating the design that produced the data at hand. G theory offers something analogous to the computation of the power of a statistical test. It helps us to check not only that the design is sufficiently dependable, but how we might later adjust the model that furnishes the measures. By separately analyzing the true and error variances, as illustrated above, G theory enables us to estimate independently the adequacy of differentiation of the objects of measurement and the precision of the measures taken of each object. These two characteristics of a measurement procedure are not distinguished when we compute an *F*-test and this makes optimization more difficult.

18.* *What can be learned from this exercise?*
   a. Even if all the intended students are tested in a performance survey (as in this example), the underlying survey design is not necessarily adequate. Sampling a sufficient number of items is just as important.
   b. The largest part of differentiation variance might be due to facets in which the objects of study are nested (like Levels in this example).
   c. The criterion-referenced dependability coefficient takes on a different meaning depending on the relative position of the cut-score and the mean of the score distribution (this was the case for the two levels).
   d. A test is not reliable in itself. It might differentiate adequately among students, but not among classes or any other facet nesting the students (Districts, Levels, etc.). A new generalizability analysis is required for each new object of measurement, to establish score reliability.
   e. A measurement procedure can have high relative reliability but low absolute reliability (low precision), and vice versa.

Reading interest

The context

Is it possible to measure the attitude to reading of six-year-old children by means of a short questionnaire filled in by the teacher? Even more interesting: Is it possible to measure a change in reading attitudes between the first and second semesters of the first grade, using this same instrument? These two questions motivated the research study described in this example.

In order to address the questions, a questionnaire was developed that comprised 10 statements describing the behavior of a child. Item 2, for instance, was "During a free time period, spontaneously picks up a book of reading exercises," while Item 5 was "Gives up reading a text after 5 minutes." Teachers were to respond to each statement with reference to a particular child, by checking the appropriate box on a four-point scale: *Never, Sometimes, Often,* and *Very frequently.* As the two example items illustrate, the behavior descriptions sometimes reflected behaviors that could be considered as evidence of a positive attitude to reading and sometimes reflected behaviors indicating a negative attitude. At the time of data entry some response scales were reversed in order to have all items rated in the same direction. The questionnaire was completed by teachers in February and again in June. The file *AttReadData.txt* contains the resulting data. There are 10,000 data points, organized as follows: data records for 250 boys are followed by data records for 250 girls, and every child has two lines of data, one for the first rating session (February) and another for the second rating session (June), each line containing 10 teacher ratings.

The questionnaire could be used as a tool for the teachers, to help them adapt their teaching strategies when necessary, to increase the motivation of each individual student. Alternatively, it could be used as a research tool, to help make comparisons between the levels of facets other than Students. For example, the Items could be considered the object of measurement, either to explore the relative frequency of different reading-related behaviors, or to estimate precisely the frequency of occurrence of each of the behaviors at this age. A G study could be used to evaluate the adequacy of the design for yielding such estimates. In addition, the two Occasions would be interesting to compare, to detect changes over time for each item separately, or for total reading interest, the expectation being that reading interest would increase over the period. Could the questionnaire yield reliable measures of change? Then again, Gender could become the focus of interest. Boys and girls are known to have different reading interests. The ability of the questionnaire to measure the gender difference in general reading interest could be considered as validation support for the instrument's content.

Questions

Overall design

1. What would be appropriate observation and estimation designs for a G study, based on the data in the file *AttReadData.txt*?

(Indicate how the designs should be declared in the EduG Workscreen, and also show how you can restrict the study to the February data only.)

2. Draw the corresponding variance partition diagram
   (for the complete design, including the June results, to match the organization of the data file).

Differentiation of individual students' attitudes

3. Could the individual students' reading interest levels be measured reliably on a relative scale in February? Did gender appreciably influence a child's level of reading interest?
4. How many questions would have been sufficient in February to keep the reliability for relative measurement above 0.80?
5. For the data collected in June, was it possible to estimate the reading interest of individual students with precision on an absolute scale, in order to, for instance, intervene appropriately in the case of a very low interest in reading? What is the corresponding standard error of absolute measurement?
6.* When separate analyses are carried out for each occasion—February and June—why does the facet O seem to have no effect in the ANOVA table (for the computations of questions 3 and 5 above)?
7. Calculate the 95% confidence interval for the absolute mean interest score of an individual student in February (i.e., for the mean over 10 items).
8. Using the February data, compute the criterion-referenced coefficient, with a λ value of 2.5 (the center of the absolute scale). Can a student located in the lower half of the scale be measured with sufficient dependability?

Change in individual students' interest levels

9.* Estimate the reliability of the change score for individual students. (*Hint*: EduG does not compute this coefficient directly, but gives the components of variance for each area of the variance partition diagram, when G, S:G, and O are declared as the differentiation face and I as the instrumentation face. The variance created by the change in reading interest of each student is contained within the O circle. The true variance of the changes is the sum of the three components in O outside of the I ellipse. The relative error variance corresponds to the three components in the intersection of O and I. Each of these components must be divided by the number of times it has been sampled for each student. The sum of these three weighted components is the relative error variance. The reliability of the change measure is the ratio of the true variance divided by the estimated total variance, that is, by the sum of the true variance and the relative error variance.)

Use of the questionnaire as a research tool

10. Find the means of each of the 10 items at the end of the year. Are the 10 underlying behaviors well differentiated by the questionnaire, both on a relative and on an absolute scale?
    (Use only the June data, to avoid combining too many sources of variance.)

11. Would it be possible, for the June rating session, to reliably rank the two genders on a relative scale? Could the mean interest levels of boys and of girls be measured with precision? Which factor contributed most to the absolute error variance?
    (Use only the June data, to avoid combining too many sources of variance.)

12. Are Occasions well differentiated over the year, using the mean of a random sample of 10 items? If not, would "fixing" the items help? And lengthening the questionnaire?

13. In fact, an improvement in the questionnaire proves necessary. Use a G-facets analysis to identify which items do not contribute well toward differentiating the two occasions.

14.* What can be learned from this exercise?

Answers

1. *What would be appropriate observation and estimation designs for a G study, based on the data in the file* AttReadData.txt?
   Students are nested in gender; hence S:G. All students are tested on two occasions; hence (S:G) × O. The questionnaire items are the same each time; hence the design is (S:G) × O × I.

   Under "Observation design reduction" in Table 5.8, the value "2," written in boldface, shows that the second rating session (June) is

Table 5.8 Declaring Facets for the Reading Interest Study

Facet	Label	Levels	Universe	Reduction (levels to exclude)
Gender	G	2	2	
Students: Gender	S:G	250	INF	
Occasions	O	2	2	**2**
Items	I	10	INF	

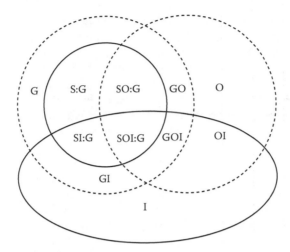

Figure 5.6 Variance partition diagram for the estimation design (S:G)OI, where S, G, O, and I represent Students, Gender, Occasions, and Items, respectively, and G and O are fixed.

temporarily excluded. Thus only the data received in February (Occasion 1) are now analyzed.

2. *Draw the corresponding variance partition diagram.*
   See Figure 5.6.

3. *Could the individual students' reading attitudes be measured reliably on a relative scale in February? Does gender appreciably influence a child's attitude to reading?*
   Because of the complexity of the data, we present the full G study results in Table 5.9. The relatively high reliability coefficients (0.87 and 0.84) confirm that good differentiation is possible at the individual student level. The relative standard error of 0.23 is small compared to the standard deviation of true scores, which reaches 0.61. The variance of true scores (0.37) is the total of two components, the variance due to the gender means (0.01) and the variance due to differences between individual scores and the respective gender means (0.36). The gender variance component is very small compared with others, reflecting the fact that in this study the overall level of reading interest was essentially the same for boys and girls—an unexpected finding.

4. *How many questions would have been sufficient in February to keep the reliability for relative measurement above 0.80?*
   Optimization shows that six items would have been enough to maintain a relative G coefficient of 0.80, but that eight items would be needed for an acceptable absolute reliability.

5. *For the data collected in June, was it possible to estimate the reading interest of individual students with precision on an absolute scale, in order to, for instance, intervene appropriately in the case of a very low interest in reading? What is the corresponding standard error of absolute measurement?*

   The value of the absolute coefficient is 0.82, which is satisfactory. The SEM (even if Gender is defined as infinite random in the estimation design) is equal to 0.256, on an absolute scale of 1–4. This seems sufficient to support educational decisions about future instruction.

6. *Calculate the 95% confidence interval for the absolute mean interest score of an individual student in February (i.e., for the mean over 10 items).*

   The standard error of the mean for absolute measurement is 0.26 (see Table 5.9). Multiply this by 1.96, giving 0.51. The 95% confidence interval is then given by the student mean ±0.51.

Table 5.9 G Study Table for Differentiating Students (February Data) (Measurement Design GS/OI)

Source of variance	Differentiation variance	Source of variance	Relative error variance	% Relative	Absolute error variance	% Absolute
G	0.0114		—		—	
S:G	0.3583		—		—	
	—	O	—		—	
	—	I	—		0.0146	21.4
	—	GO	—		—	
	—	GI	0.0002	0.4	0.0002	0.3
	—	SO:G	—		—	
	—	SI:G	0.0534	99.6	0.0534	78.3
	—	OI	—		—	
	—	GOI	—		—	
	—	SOI:G	—		—	
Sum of variances	0.3697		0.0536	100%	0.0682	100%
Standard deviation	0.6080		Relative SE: 0.2316		Absolute SE: 0.2612	
Coef_G relative	0.87					
Coef_G absolute	0.84					

7.* *When separate analyses are carried out for each occasion—February and June—why does the facet O seem to have no effect in the ANOVA table (for the computations of questions 3 and 5 above)?*
This is simply because the data analyzed relate to one occasion only—the other having been excluded, meaning that there can be no variance between occasions.

8. *Using the February data, compute the criterion-referenced coefficient, with a λ value of 2.5 (the center of the absolute scale). Can a student located in the lower half of the scale be measured with sufficient dependability?*
The estimate of the criterion-referenced dependability coefficient is 0.92. This shows that a student with low reading interest can be identified with precision already in February, without having to make comparisons with other students.

9.* *Estimate the reliability of the change score for individual students.*
A new G study taking into account the data of both semesters gives rather similar results, but allows a more detailed analysis.
   The totals of the three columns of Table 5.10 enable us in the usual way to compute the proportion of the total variance that is due to the true variance between students, when Genders and Occasions have been pooled. The error variance is then due to the interaction with Items (for the relative G coefficient). To this error must be added the variance between items, if an absolute coefficient is computed.
   Less direct is the computation of a within-student G study of change scores. The components G and S:G express the between-students variance, if their annual means are considered. This effect should be discarded from the true variance, if only individual progress is of interest. The components of the true variance for individual change scores are the sum of the variance components for Occasions (O), Gender by Occasions (GO), and Students (within Gender) by Occasions (SO:G), that is, $0.000442 + (0.0000) + 0.028906 = 0.029348$. (Note that the estimated variance component for GO was negative and consequently replaced by zero.) [It is also interesting to note that the standard errors reported here in Table 5.10 (uncorrected for facet size) differ only in the third decimal place from those reported above in Table 5.9.]
   The corresponding relative error variance is the sum of the variance components for Occasions by Items (OI), Gender by Occasions by Items (GOI), and Students (within Gender) by Occasions by Items (SOI:G), each component being divided by 10, the number of items (behaviors) rated for each student. These

Table 5.10 G Study Table for Individual Scores (Measurement Design SOG/I)

Source of variance	Differentiation variance	Source of variance	Relative error variance	% Relative	Absolute error variance	% Absolute
G	0.010196		—		—	
S:G	0.298476		—		—	
O	0.000442		—		—	
	—	I	—		0.013421	20.0
GO	(0.0000)		—		—	
	—	GI	0.000220	0.4	0.000220	0.3
SO:G	0.028906		—		—	
	—	SI:G	0.037605	70.2	0.037605	56.1
	—	OI	0.000140	0.3	0.000140	0.2
	—	GOI	0.000010	0.0	0.000010	0.0
	—	SOI:G	0.015596	29.1	0.015596	23.3
Sum of variances	0.338020		0.053571	100%	0.066992	100%
Standard deviation	0.581395		0.231455		0.258829	
Coef_G relative		0.86				
Coef_G absolute		0.83				

values appear among others in the column under Relative error variance: $0.000140 + 0.000010 + 0.015596 = 0.015746$.

The estimated total variance (a G coefficient is a ratio of estimates and not of observed variances) is the sum of the true and the error variance: that is, $0.029348 + 0.015746 = 0.045094$, or 0.045 rounded. The proportion of true variance in the total variance is $0.029348/0.045094 = 0.65$. This index of reliability is far from sufficient. To raise it to at least 0.80, the denominator of the ratio would have to be a maximum of 0.036685, and the relative error variance would have to be a maximum of $0.036685 - 0.029348$, or 0.007337. This could be achieved by multiplying by $0.015746/0.007337$, or 2.17, the number of items rated by the teachers for each student, giving 22 items in total. But this might not be workable in practice.

10. *Find the means of each of the 10 items at the end of the year. Are the 10 underlying behaviors well differentiated by the questionnaire, both on a relative and on an absolute scale?*
    The means for the 10 items are shown in Table 5.11.

Table 5.11 Item Mean Scores at the End of the Year

Items	1	2	3	4	5	6	7	8	9	10
Means	3.32	3.00	3.06	2.71	3.06	3.06	3.75	3.69	2.74	3.37

With Items on the differentiation face and the three other facets on the instrumentation face, the reliability coefficients for relative and absolute measurement would both have a value of 0.99 rounded. The situation could hardly be better. This result can be explained (1) by the fact that the variance between the items is sizeable, but mostly (2) by the large number of students assessed, which reduces the contribution of the student-by-item interaction to the error variance. The largest difference in item means is more than one point on the four-point scale. The Items facet is therefore an important source of variance, which contributes to the questionnaire's ability to differentiate the various student behaviors as frequent or infrequent, but will be detrimental should the aim rather be to compare individuals or genders or occasions. This particular error variance, however, affects only the precision of the estimation of the *absolute* value of each behavior, not their relative standing in frequency terms. This explains the results for question 11.

11. *Would it be possible, for the June rating session, to reliably rank the two genders on a relative scale? Could the mean interest levels of boys and of girls be measured with precision? Which factor contributed most to the absolute error variance?*
    The value of the relative reliability coefficient for Gender for the June rating session is 0.84, which is quite satisfactory. Absolute differentiation, on the other hand, gives a G coefficient of only 0.384. The difference is striking. It can be explained on the basis of Table 5.10, which shows that (when the Genders are differentiated) 88% of the absolute error variance comes from the Items effect. This is why relative reliability can be good, while the precision of the measures themselves can be inadequate. The high between-item variance, in terms of differences between Item means, increases the size of the denominator of Coef_G absolute.

12. *Are Occasions well differentiated over the year, using the mean of a random sample of 10 items? If not, would "fixing" the items help? And lengthening the questionnaire?*
    EduG shows that the measure of change between occasions is not reliable, even for the mean of all observations in February compared to the mean of all observations in June, if we treat Items as a random facet, that is, if we accept that a new item sample can be drawn on the

second occasion of testing. The relative coefficient is then only 0.66, while the absolute coefficient is practically zero (0.03). If, on the other hand, we use the same 10 items each time, essentially treating the Items facet as fixed, then the relative G coefficient reaches 0.87. Even then, however, the absolute coefficient remains at 0.433, essentially because of the high between-student variance. Increasing the number of items would not be enough to improve matters, if we continue randomly sampling the items. An optimization study shows that 35 questions would be necessary to bring the reliability of relative measures to a satisfactory level (the absolute reliability remaining lower than 0.10). But the task of rating students against so many behavior statements would be too heavy a burden for busy teachers.

13. *In fact, an improvement in the questionnaire proves necessary. Use a G-facets analysis to identify which items do not contribute well toward differentiating the two occasions.*

Table 5.12 presents the results of the G-facets analysis. If the relative coefficient increases markedly when an item is dropped from the set, this shows that this item might not belong to the same dimension as the others. It should be reviewed for possible flaws or misclassification. Items 8 and 9 might be worth looking at in this case, since when each is excluded from the data set the reliability increases from an average of around 0.6 to around 0.8. (The Items universe was supposed randomly sampled, so that simply dropping these two items to increase reliability would not be justified—unless a new universe could be clearly defined that did not contain items of this general type, if such a "general type" exists.)

Table 5.12 Items G-Facets Analysis

Facet	Level	Coef_G relative	Coef_G absolute
I	1	0.61	0.02
	2	0.52	0.02
	3	0.65	0.03
	4	0.50	0.02
	5	0.59	0.02
	6	0.44	0.01
	7	0.65	0.04
	8	0.81	0.07
	9	0.79	0.05
	10	0.43	0.01

14.* *What can be learned from this exercise?*

Obviously, generalizability analyses can be carried out with subsets of the available data, as illustrated earlier when the February and the June questionnaire data sets were separately analyzed. We have also learned that it is possible to combine or to separate individual components of variance, as in the answer to question 9. When the reliability of individual progress was being considered, the G and S:G facets were discarded, in order to make the design apply to each individual student separately. The only facets specifically identified then were O for the face of differentiation and I for the face of instrumentation. G and S:G played the role of "hidden facets." It is often useful to work like this with the complete design, using all of the available data in the G study analysis in order to estimate each source of variance separately, and then to arrive later at particular estimates by combining the variance components like building blocks.

Laboratory-based examination

The context

In a graduate Chemistry course, the end-of-year assessment required students to carry out two laboratory experiments, each focusing on a different topic chosen at random from the 10-chapter set text covered during the year. For practical reasons, the two topics were the same for all students in any particular year, but varied from one year to another. Students were randomly assigned to one or other of three parallel groups for the examination itself, each group comprising approximately 30 students. The groups were observed and assessed by different examiners, as it was not possible for one examiner to examine all the students in the half-day available.

Each examiner observed each student in the relevant group individually conducting the experiments, one experiment in the morning and the other in the afternoon. The order in which students were assessed was randomly determined. The examiners used a checklist to rate each student against a number of procedures as the experiment progressed, scoring each procedure 1 for correct, 0 for incorrect, and 0.5 for an unclear outcome. While some of the procedures were unique to the particular experiments in each year, nine were common to all the experiments. Since the experiments varied in inherent difficulty an adjustment was applied to the resulting total scores to bring mean scores for all experiments into line. A consequence was that each year the examination results over the two assigned experiments would have the same mean.

The Chemistry faculty were keen to evaluate the assessment system that they were using, and so carried out a G study, using two years' worth of data relating to the nine procedures that were common to all experiments. To equalize the number of observations, 28 students were drawn at random from within each of the six groups (three per year), this being the minimum number per group in the period. The data are contained in the file *LabData.txt*. Each student has two records in the file, each record listing the nine procedure ratings for one or other experiment (topic). The student records for one year are followed by those of the other year, and within each year the records are sorted by laboratory group.

The facets involved in the study are Years, Groups, Students, Topics, and Procedures. Since each group had a different examiner, the Groups facet is actually a confounding of Groups and Examiners (i.e., the two effects cannot be separated). Students and examiners differed from year to year, as did topics. Thus, Students and Topics are nested within Years. This example differs from the previous four, in that readers are invited to check the answer to each question before moving on to the next, since there is a logical investigative thread running through the question sequence.

Questions

Global design

1.* What are the observation and estimation designs for the proposed G study?

   In the EduG Workscreen, list the facets in the order they are mentioned above, since the declaration must correspond to the data structure in the data file LabData.txt.

   (*Hint*: Start with the facet Years and determine which facets it is nesting. Do the same with the facets Groups and Students. Identify these facets in a way that expresses their relationships. Then examine the facet Topics and find its relationship with each of the previous three facets. Identify it accordingly. Finally, indicate the relationship between the facet Procedures and each of the other four facets. Remember that two facets that are not nested one within the other are crossed. Consider all facets as completely random, with the exception of Topics, for which the sampling is random finite.)

2.* Draw the diagram for the estimation design.

   [*Hint*: Follow the order indicated above. Draw a large circle for Years, and within it draw circles to represent the other facets and their inter-relationships (circle included if nested, intersecting if crossed)].

3. How many variance components contribute to the absolute error variance for student measurement?

Examination reliability

4. What is the measurement design for exploring the reliability of student scores?
5. What are the values of the two reliability coefficients for student scores, relative and absolute? Which one is most relevant for such an examination? Is it high enough?

Optimization strategies

6. What would be the value of the reliability coefficients should the number of experiments each student was required to carry out be greater than two—say three, four, five, or six?
7. What would be the value of the coefficients should the number of rated procedures be 12 or 18?

Unexpected sources of error variance

8. Is there a problem with one or other of the four topics, or with one or other of the six year groups? To find out, compare the mean results for each topic and each group in each year to the standard error of these means.
9. Are there some procedures that do not seem to have been tapping the same dimension of laboratory performance? Use a G-facets analysis to find out.
10.* Try to carry out a Topics facets analysis. Why does EduG turn down your request?

Analysis of the absolute error variance

11. Compute (*to six decimal places*) the components of absolute error variance for the individual student examination scores (which are averages across 18 procedure scores, nine from the morning topic and nine from the afternoon topic in each year). What are the main sources of sampling error?
12. Compute (*to six decimal places*) the components of absolute error variance for the individual topic mean scores (which are averages across nine procedure scores). What are the remaining main sources of sampling error? What is the reliability of these scores on an absolute scale?

Criterion-referenced dependability

13. Was the examination able to measure reliably the distance of an individual student from a pass score of 75%, or of 66%?

Precision of the examination scores

14. What is the value of the standard error of absolute measurement for the students' examination scores?

Conclusion

15.* What do you conclude from the analysis results about the quality of the chemistry examination?
16.* What can be learned from this exercise?

Answers

1.* *What are the observation and estimation designs for the proposed G study?*
Students are nested within Groups, and Groups are nested within Years. Years, Groups, and Students are therefore declared as Y, G:Y, and S:G:Y, respectively. Topics vary from year to year, so Topics are also nested within Years, hence T:Y. The nine Procedures for which scores are available in the data set were the same for all experiments, and so this facet is crossed with all other facets, being declared, therefore, simply as P (see Table 5.13).

All facets are considered infinite, with the exception of Topics, since topics are drawn from the limited number included in the 10-chapter text that underpins the course.

2.* *Draw the diagram for the estimation design.*
See Figure 5.7.

Table 5.13 Declaring Facets for the Chemistry Laboratory G Study

Facet	Label	Levels	Universe
Year	Y	2	INF
Groups in years	G:Y	3	INF
Students in groups in years	S:G:Y	28	INF
Topics in years	T:Y	2	10
Procedures	P	9	INF

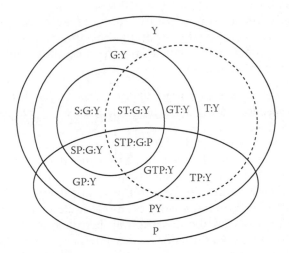

Figure 5.7 Variance partition diagram for the estimation design (S:G:Y)(T:Y)P, where S, G, T, Y, and P represent Students, Groups, Years, Topics, and Procedures, respectively, and T is fixed.

3. *How many variance components contribute to the absolute error variance for students?*
   A total of 13 different variance components feature in the design. Nine of them contribute to the relative error variance for student measurement: all the interaction effects between S, G, and Y on the one hand and T and P on the other (since S is nested within G, and G within Y, the components Y and G:Y join S:G:Y to comprise the differentiation variance). There are 10 components in the absolute error variance, the P component being added to the previous nine.

4. *What is the measurement design for exploring the reliability of student scores?*
   Each student score is the mean of 18 values, nine for each of the two experiments (morning and afternoon). The objects of study are the Students, along with the facets in which they are nested, that is, Groups and Years. These three facets form the face of differentiation. The instruments of observation are the Topics and the Procedures, forming the face of instrumentation. So the measurement design is declared to EduG as YGS/TP (or equivalently as GSY/TP, GSY/PT, etc.). The true variance corresponds to the three areas within the differentiation face and outside of the instrumentation face, that is, S:G:Y, G:Y, and Y. The elements of the relative error variance are the nine areas belonging both to the circles of the differentiation facets (Y, G, and S) and to those of the generalization facets (T and P). They are: ST:G:Y, GT:Y, T:Y, SP:G:Y, STP:G:Y, GTP:Y, GP:Y, TP:Y, and PY. The faculty regret having only nine

common procedures rated in the study, because the components of relative error variance are all divided by the number of procedures rated per student, so that more common procedures would have meant lower measurement error.

5. *What are the values of the two reliability coefficients for student scores, relative and absolute? Which one is most relevant for such an examination? Is it high enough?*

   The relative coefficient is 0.24, and the absolute coefficient is 0.23. This last value is what should be considered for this particular examination, whose function is not to rank students. The faculty would have been dismayed with this result. The absolute reliability is much too low: three-quarters of the estimated total variance of the student means was due to sampling fluctuations that ought not to have affected the examination scores. Furthermore, the G study table shows that nearly half of the true variance is due to differences between the groups. These differences could reflect differences in the caliber of the students in the various groups. But as the groups were formed at random they are more likely due to marking differences between the different examiners. Clearly, attention turned to how to change the examination in the future, in order to improve its reliability for student assessment. One possibility would be to require students to perform more than two experiments. Another might be to try to ensure that the number of procedures common to the experiments was higher than nine. A third would be to have more than one examiner examine each student, or, to save time and costs, to ensure through prior training that the examiners were all applying similar rating standards.

6. *What would be the value of the reliability coefficients should the number of experiments each student was required to carry out be greater than two— three, four, five or six, for example?*

   An optimization study shows that doubling the number of experiments, from two to four, would increase the reliability for absolute measurement from 0.23 to 0.37, still far too low. With six experiments the absolute coefficient would still only reach 0.47, and the workload for students and faculty would anyway be too heavy. This strategy— adding more experiments—is therefore not a useful option. So what about increasing the number of procedures common to all experiments, if that might be possible?

7. *What would be the value of the coefficients should the number of rated procedures be 12 or 18?*

Unhappily, a second optimization study reveals that increasing the number of common procedures would not improve the reliability of the examination as a whole. With 12 common procedures the absolute coefficient would be 0.26, while 18 common procedures would only result in a very slight further increase to 0.30. So even doubling the number of common procedures would only increase the reliability from the current 0.23 to 0.30.

The obvious optimization strategies having failed, what could the faculty explore next? Perhaps one or two of the examiners were out of line with the others in their rating standards? Even though groups and examiners were confounded, it might nevertheless be informative to look at the group by topic scores in each year, to see whether anything unusual might emerge.

8. *Is there a problem with one or other of the four topics, or with one or other of the six year groups? To find out, compare the mean results for each topic and each group in each year to the standard error of these means.*
An EduG request for Means for GT:Y provides the result shown in Table 5.14. Group 3 examiner (although different from one year to the next) seems to have been more severe than the others in each of the two years. But could this be a normal and inevitable random fluctuation?

A G study differentiating these means (measurement design YTG/SP) shows that the absolute standard error for the group averages is equal to 0.055, and the overall mean score is 0.871. The 68% confidence interval around this estimated mean spans 0.816–0.926. Assuming a Normal distribution for group means, two-thirds of these 12 means should lie between these limits. We see, in fact, that eight of the 12 means do fall within the confidence interval, with two above and two below. The distribution could not be more Normal (notwithstanding the small number of observations!). Thus random fluctuations could very easily create variations of this size

Table 5.14 Mean Scores for Each Topic and Group in Each Year

| | Year 1 | | Year 2 | |
| | 0.855 | | 0.887 | |
Year group	Topic 1	Topic 2	Topic 3	Topic 4
Group 1	0.908	0.847	0.915	0.865
Group 2	0.938	0.833	0.907	0.950
Group 3	0.819	0.782	0.891	0.794
All groups	0.889	0.821	0.904	0.870

between the group means. Other reasons for the low examination reliability need to be looked for. The values in Table 5.14 are interesting, however, in showing that the level of student success is consistently high.

9. *Are there some procedures that do not seem to have been tapping the same dimension of laboratory performance? Use a G-facets analysis to find out.*
The results of the facet analysis for Procedures are given in Table 5.15, which shows that all the reliability coefficients are of the same order, with the exception of the lower 0.14, which arose when procedure 4 was excluded from the analysis. So, there were no particular procedures that were serving to reduce examination reliability by their inclusion. What, then, of Topics?

10.* *Try to carry out a Topics facets analysis. Why does EduG turn down your request?*
For the Procedures facet, which was crossed with all the differentiation facets, it was possible to drop one level at a time and to compute the reliability of the mean of the remaining procedures. But one cannot drop a single level of the facet Topics, because no level, or topic, was the same in both years. Topics, being nested in Years, has no meaningful level 1 or level 2. A G-facets analysis for this facet would, therefore, make no sense, and this is why EduG rejects the request for one.
   There are no solutions to the reliability problem so far. So let us analyze further the sources of sampling variance.

Table 5.15 Reliability after Deletion of Each Procedure Separately

Procedure	Coef_G absolute
1	0.22
2	0.24
3	0.24
4	0.14
5	0.24
6	0.22
7	0.24
8	0.23
9	0.20

11. *Compute (to six decimal places) the components of absolute error variance for the individual student examination scores (which are averages across 18 procedure scores, nine from the morning topic and nine from the afternoon topic in each year). What are the main sources of sampling error?*

    A G study with YGS/TP as measurement design shows that just three variance components account for 82% of the total absolute error variance. They are, in order of importance:

    a. ST:G:Y, the Student-by-Topic interaction, which reflects the fact that some students are more at ease with one topic than with the other, and vice versa. This is unavoidable, but regrettably important, accounting for 37% of the total absolute error variance of the examination scores.

    b. STP:G:Y, the highest-order interaction, between Students, Topics, and Procedures. This indicates that individual students did not carry out the various procedures equally well in both experiments. The component accounts for 33% of the total, and would be difficult, if not impossible, to control (without more laboratory practice for the students beforehand, to improve their performance consistency).

    c. SP:G:Y, the interaction between Students and Procedures, represents 11% of the total error variance for absolute measurement. This interaction effect reflects the fact that different students do better or worse on different procedures, irrespective of the particular experiment being carried out (again, more prior laboratory experience might address this issue).

    The choice of topic clearly has an important influence on a student's examination result. The component ST:G:Y, in particular, contributes to the error variance when the overall examination mean is calculated. If, however, we look instead at the students' mean scores for each individual experiment, then this source of variance will disappear and score reliability should improve.

12. *Compute (to six decimal places) the components of absolute error variance for the individual topic mean scores (which are averages across nine procedure scores). What are the remaining main sources of sampling error? What is the reliability of these scores on an absolute scale?*

    A G study with YGST/P as measurement design shows that three components out of the 13 account for 89% of the total absolute error variance. They are, in order of importance:

    a. STP:G:Y, the third-order interaction, which accounts for 67% of the absolute error variance.

    b. TP:Y, which reflects the fact that some procedures are better executed for one topic than the other within each year (a 12% contribution).

    c. SP:G:Y, which shows that some procedures are easier for some students than for others and accounts for 10% of the total absolute error variance.

Score reliability for topic means reaches 0.51, which is still way too low. Yet it is difficult to see how the important interaction effects might be reduced, simply by changing the examination format. Let us now look at other aspects of the examination.

13. *Was the examination able to measure reliably the distance of an individual student from a pass score of 75%, or of 66%?*

    With a passing score of 75%, the value of the criterion-referenced dependability coefficient, $\Phi(0.75)$, is 0.64. If the cut-score is lowered to 66%, then the coefficient, $\Phi(0.66)$, increases to 0.83, an adequate value. A dependability coefficient of 0.80 is obtained for a cut-score of 68%. All is not lost, then. The examination can reliably be used to separate students into two groups, either side of a mean score of 68%. But what, finally, is the confidence interval for an individual student mean score (mean of 18 procedure scores)?

14. *What is the value of the standard error of absolute measurement for the students' examination scores?*

    EduG shows that the value of the absolute SEM is equal to 0.096. The margin of error, for a 95% confidence interval, is then ±0.188.

15.* *What do you conclude from these analysis results about the quality of the chemistry examination?*

    The examination was not precise enough to rank students reliably, and yet the absolute error variance is of moderate size, with a margin of error of ±0.188. The proportion of true score variance in the total observed score variance is small, principally because the true score variance is itself low. The mean of the examination score distribution is equal to 0.871 on the 0–1 scale, leaving little room for variation among students. So the situation was not as bad as it seemed at first sight. Most of the students did very well on the examination, and their score distribution was in consequence very narrow. This is the situation that Benjamin Bloom dreamed of when he wrote *Learning for mastery* (Bloom, 1968). As there was little between-student variation in scores, the numerators in the relative and absolute coefficients were small, resulting in low coefficient values. The examination, on the other hand, could reliably be used to separate good performers from others by using a cut-score of 68%. But a new question then arises: Could a cut-score of 68% be justified in any way other than for its ability to throw a positive light on the reliability of this particular examination?

16.* *What can be learned from this exercise?*
The general intent of this exercise was to review and summarize the techniques presented earlier, to show how sources of error that affect the quality of our measurement procedures can be detected and possibly controlled. When the reliability of the particular examination studied proved inadequate, we showed how an optimization analysis could be used to explore the most obvious and the potentially most effective strategies for addressing the problem. When standard optimization analysis failed to suggest a solution, other ways of investigating the data were pursued, in the same way that a doctor might work systematically toward a medical diagnosis. In other words, the message from this exercise is that a G study cannot always usefully follow a rigid pattern of questioning, but should be guided by a systematic line of attack. With each new piece of information acquired, the investigator can usually see which way to proceed until a clear improvement strategy becomes evident, or it is confirmed that none exists. These are the characteristics of the investigative methodology that this exercise was intended to illustrate.

chapter six

Current developments and future possibilities

We begin this chapter with a brief overview of the evolution in thinking that has characterized G theory development over the last 40 years or so, before moving on to describe how, and how widely, G theory has been applied in practice over the last decade. We then highlight the new focus on effect size and SEM estimation, before attempting to identify possibilities for further development of G theory, including its integration within a single unified theory of measurement.

Origin and evolution

G theory derives from correlation-based classical psychometrics, whose most characteristic feature is the absence of any distinction between sample and population. The great achievement of Cronbach and his associates in developing G theory was in adopting the ANOVA as the underpinning foundation of their psychometric formulas. Their seminal article (Cronbach et al., 1963) brought together a multitude of previously proposed formulas within a coherent framework of variance partition, in which different factors (facets), as sources of sampling variance, contribute to measurement error: items, markers, occasions of testing, and so on. In its earliest conception the aim of G theory was to produce the best measurement instruments for use in student evaluation. In those early days, students were almost exclusively the object of measurement, and while the dependability of test results for individual students was clearly the primary focus, students as well as items and other aspects of measurement were treated as random factors for the purposes of score reliability estimation. In other words, $E\hat{\rho}^2$, the original "generalizability coefficient," was defined in terms of a completely random model. G theory was adapted to mixed-model ANOVA by Brennan (1986, 2001). But this was essentially to describe the effect on error variance of the random sampling of items or conditions drawn from finite universes. Differentiation facets remained defined as purely random.

In their articles of 1976 and 1981, Cardinet and Allal noted that in the ANOVA model all factors have the same status, and the estimation of variance components proceeds in the same way irrespective of whether, for example, it is differences between items, differences between people, or differences between some other types of entity that are of most interest. It is the factor sampling scheme (i.e., the fixed, finite random or random status of each factor) that is of primary importance in component estimation, as we explained in Chapter 2. The intention behind the symmetry articles was to encourage the application of G theory to objects of study other than students and to measuring instruments (or procedures) other than tests, questionnaires, and examinations. Acceptance of the symmetry principle brings new challenges, however. This is because *any* factor can now be considered the object of measurement, with the consequence that this is no longer necessarily infinite random. It can in certain cases be finite random, or even fixed. But the statistical model underpinning G theory is affected both conceptually and technically by fixing the differentiation facet, on the one hand in terms of the very meaning of the generalizability model, and on the other in terms of estimation and interpretation of generalizability parameters.

To address this crucial issue, let us briefly consider the two opposing cases: infinite random or fixed differentiation facet. Think of a very simple example in educational research, that is, measuring the results of teaching, with learning objectives (L) as the differentiation facet and students (S) as the instrumentation facet (the observation and measurement designs are LS and L/S, respectively). Imagine also that the G coefficient for the resulting set of data is high, say 0.90. Given this, what conclusions can we draw, depending on whether the differentiation facet is infinite random or fixed?

In the first case—an infinite random differentiation facet—the results confirm that measurement of differences in success across the various learning objectives would be reliable no matter which particular individuals were involved in the study (students randomly selected from within a given target population) *and* no matter what learning objectives were actually considered (similarly randomly selected from within a target domain). We can say that the study permits a double generalization, extending to the domain of learning objectives as much as to the student population. In this sense, the elements measured can be considered *generic* (any, interchangeable, anonymous), since the study outcome does not exclusively relate to the learning objectives that are actually featured, but extends to every random sample of learning objectives belonging to the same domain. The standard error also has a generic character, in the sense that it indicates the "expected" error associated with any learning objective, whether observed or not.

In the second case—a fixed differentiation facet—one would conclude that the measurement would be reliable only for the *particular* learning objectives considered in the study. In this case generalization is possible, but it applies exclusively to the elements belonging to the instrumentation facet, as initially defined. Here, the object of study has a *specific* character (identifiable, singular), because measurement reliability is assured only for the learning objectives actually observed. The standard error is therefore no longer a generic measurement applying to no matter what object of study, but becomes a parameter specifically associated with the elements with which it was calculated. We came across this issue in several of the example applications in Chapter 4: Example 3, where differences between fixed student Groups were of interest; Example 5, where the relative effectiveness of particular methods of teaching writing were under evaluation; and Example 6, where attitude change over a given time period was in focus. For further applications consult http://www. irdp.ch/edumetrie/documents/AutresDomainesD%27Applicat.pdf.

In experimental design applications, in fields such as agriculture and industry, "fixed" objects of study are the rule. Mixed model ANOVA was essentially developed in this context (Fisher, 1955). When measurements are collected during a designed experiment, whatever the field of application, the focus of interest is typically the manipulated independent variable, whose levels are usually fixed. In medical research, for example, pharmaceutical X might be compared for effectiveness with pharmaceuticals Y and Z. In psychiatry one type of therapy might be compared with another. In educational research, teaching Method A might be compared with teaching Method B, as in the illustrative comparative study discussed in Chapters 1–3, with students evaluated at the start and at the end of a module.

It is appropriate for G theory to be used in experimental design applications. But we have to recognize that by modifying the sampling status of the object of study, we depart from the theoretical framework initially adopted for G theory. This raises a technical issue in that, as we have noted in Chapter 2, the coefficient ω^2 now needs to be calculated rather than the usual ρ^2. The use of ω^2 for measuring effect size in this type of situation has been recommended by, for example, Tatsuoka (1993), who identified a multitude of impact coefficients of this type, along with their relative merits. The only inconvenience mentioned for ω^2 by Tatsuoka is calculation complexity. With EduG this problem is resolved, since the appropriate G coefficients are provided without users needing to be aware of their exact nature (ω^2, ρ^2, or values in between for finite random facets). Coef_G, the coefficient computed by EduG, is a generic reliability index that simply estimates this proportion of universe variance in the estimated total variance in accordance with the given estimation design.

EduG's general calculation procedure retains meaning even if the object of study comprises a single level (e.g., a single curriculum objective or a single moment of observation). The formulas routinely used to define error variance can accommodate this type of calculation. In the case of a fixed differentiation facet they give exactly the right result when the appropriate algorithm is employed. This is illustrated in Example 4 in Chapter 4, where an observation design reduction was employed to calculate the various G parameters for one particular student group among those initially compared. Thus, even in the absence of differentiation variance (which makes calculation of a G coefficient impossible), the logic of the procedure remains meaningful when researchers aim to improve the precision of their measurements—looking for a standard error that does not exceed a given value, or for a confidence interval falling within certain limits. EduG can accommodate cases with a single-level object of study, or even cases where differentiation does not feature at all.

Recent G theory applications

A review of the academic literature that has accumulated over recent years reveals that G theory has been most frequently applied in the fields of educational research and measurement theory. But medicine and sports follow, the principle of symmetry here being most frequently applied in evaluating measuring procedures. Social work studies, along with business and economics research, can also be mentioned.

We can cite, for example, Auewarakul, Downing, Praditsuwan, and Jaturatamrong (2005), on the reliability of structured clinical examinations in medical education in Thailand; Boulet, McKinley, Whelan, and Hambleton (2003), on quality assurance methods for performance-based assessments in the United States; Boulet, Rebbecchi, Denton, McKinley, and Whelan (2004), on the reliability of the assessment of the written communication skills of medical school graduates in the United States; Burch, Norman, Schmidt, and van der Vleuten (2008), who used univariate and multivariate G theory to investigate the reliability of specialist medical certification examinations in South Africa; Christophersen, Helseth, and Lund (2008), on the psychometric properties of a quality-of-life questionnaire for adolescents in Norway; Denner, Salzman, and Bangert (2001), on the use of work samples in teacher education in the United States; Heck, Johnsrud, and Rosser (2000), on the evaluation of the leadership effectiveness of academic deans in the United States; Heijne-Penninga, Kuks, Schönrock-Adema, Snijders, and Cohen-Schotanus (2008), on the equivalence of different forms of written assessment in the health sciences in the

Netherlands; Johnson (2008), on the reliability of attainment estimation in a national assessment program in Scotland; Kreiter, Gordon, Elliott, and Ferguson (1999), on the reliability of expert ratings of students' performances on computerized clinical simulations in the United States; Kreiter, Yin, Solow, and Brennan (2004), on the reliability of medical school admission interviews in the United States; Magzoub, Schmidt, Dolmans, and Abdelhameed (1998), on the reliability of observation-based peer group assessment in community-based medical education in Sudan; Marcoulides and Heck (1992), on the assessment of leadership effectiveness in business education in the United States; Shiell and Hawe (2006), on the assessment of "willingness to pay" in health economics in Australia; Solomon and Ferenchick (2004), on sources of measurement error in an ECG examination in the United States.

In almost all these published studies, the underlying measurement design has been entirely random, or mixed with an infinite random object of study. But in a conventional experimental situation, as noted earlier, the object of study is generally "fixed," comprising two or more specifically selected qualitatively or quantitatively different levels. These might be different styles of teaching, different amounts of fertilizer, different examination formats, different numbers of hours of therapy, and so on. In this type of situation, it is precisely the causal effect of the fixed independent variable that must be optimally measured. The conditions under which the study is conducted are a source of sampling fluctuation, to be investigated and controlled in order to establish the margin of uncertainty associated with measurement of the fixed effect. And as usual the optimization plan must show how to reduce the SEM, if this is possible. But it is no longer meaningful to investigate the issue of "reliability" as though it were a general property of the measurement procedure, because there is no longer any need for an inference drawn on the basis of a sample of observed levels representing the differentiation facet. The "true variance" results solely from the researcher's choice of levels for the object of study. The G coefficient is now to be considered rather as an indicator of effect size, that is to say of the relative impact that the variation under investigation has on the observed results, whether these be test scores, observer ratings, grain yields, interviewer perceptions, or whatever. It still measures the importance of "true" or "valid" variance (genuine differences between the objects measured) relative to the total variance (true variance plus error variance), but the true variance no longer results from *sampling* the independent variable.

The use of the G coefficient as an effect size indicator is entirely in line with current reporting requirements for academic and scientific research. Criterion levels of statistical significance are essentially arbitrary choices, and levels of significance reached in practice are a direct function of the number of observations with which significance tests are calculated. A significance test is a "weak" argument in scientific terms, admitted Fisher (1955). Others go as far as to say that it is the most primitive type of analysis in statistics, because it comments only negatively on an artificial hypothesis formulated specifically to be rejected. For instance, Anscombe (1956, pp. 24–27) writes: "Tests of the null hypothesis that there is no difference between certain treatments are often made in the analysis of agricultural and industrial experiments in which alternative methods or processes are compared. Such tests are … totally irrelevant. What are needed are estimates of magnitudes of effects, with standard errors." While it was once possible to respond to the requirements of scientific journals by giving the effect size, plus a margin of error, these requirements were too quickly replaced by that of the probability level for the obtained result. Numerous methodologists regret this. The move back to effect size calculation is welcome, and G theory can support this.

On a related issue, several researchers have called for the mechanical application of significance tests to give way to a more thoughtful type of data exploration. Cronbach, for example, saw one of the most important outcomes of a G study as the quantification of the relative importance of the sources of variance affecting the variable under investigation. This is why he proposed that crossed factorial designs be implemented as a first stage in research, so that any interaction effects might be revealed, effects that are often more important or more instructive in practice than main effects. When interaction effects are shown to be small or nonexistent, D study designs can exploit this by substituting facet crossing by the more cost-efficient facet nesting. EduG was designed exactly from this perspective, since it allows a data set to be divided into nonoverlapping subsets and a comparison made of the parameters obtained in each. New working hypotheses have a chance to appear during this systematic data exploration.

This is an approach that is also recommended by other behavioral scientists, such as Gigerenzer (1993). Instead of losing sleep deciding rejection probabilities for the null hypothesis, researchers, he observed, should (1) focus on controlling the importance of the effects they investigate and (2) try to minimize experimental error in their measurement tools by alternating pilot studies with subsequent modifications. One sees the parallelism here with G and D studies, except that this time the procedure is applied to the fixed facets of formal experimental designs. Different checks for data heterogeneity can be readily implemented with EduG. We note also that, with his second recommendation, Gigerenzer agrees with Cronbach, since the latter asked that more importance be paid to the SEM than to the G coefficient in educational measurement applications (Cronbach & Shavelson, 2004). The SEM has a clear meaning, and should be the first consideration in any G study, with D studies employed to identify sampling strategies that would reduce the SEM to any desired level, if that level is achievable in practice. From this perspective, the optimization procedure could consist of exploring the effect of manipulation of one or more instrumentation facets on the SEM, rather than, or as well as, on the G coefficient.

G theory also has the potential to validate universe definitions. The procedures we have outlined in this volume, notably EduG's facets analysis, may lead to rejection of data or analyses that turn out not to correspond to the populations, universes, or relationships assumed at the start. But the reverse can also hold, in that it could be the theoretical hypotheses rather than the data that are put in doubt by analysis results. In this kind of application definitions of populations and universes would change on the basis of a rather heuristic procedure. When facet analysis shows that certain items, or certain subjects, behave differently from the rest, further analysis of item or subject subsets can reveal the existence of heterogeneous populations or even of distinct factors, as we have illustrated in several of the application examples in Chapter 5. It would then be appropriate to create narrower, more homogeneous, item domains or subject populations. Theoretical concepts can thus be evaluated in terms of the extent to which they are sufficiently measurable to be judged scientifically consistent, bearing in mind that a balance always needs to be found between desired measurement precision and the size of the domain to which results can be generalized.

Although G theory has been used quite widely in recent years, the absence of flexible user-friendly software has nevertheless hindered further progress. Bell (1985) identified the software problem over 20 years ago, and the problem persists to some extent today—see, for example, Heck et al. (2000, p. 682) and Borsboom (2006, p. 435). There is no doubt that once the existence of EduG becomes widely known, not only will the number of G theory applications quickly multiply, but so, too, will the number of studies in entirely new fields of application. This same outcome can be expected once commercial software packages include a user-friendly G theory module.

G theory and item response theory

It would not be appropriate to end a volume on G theory without briefly addressing its relationship with the other modern measurement theory, item response theory (IRT). IRT (Lord, 1952; Lord & Novick, 1968; Rasch, 1960; Wright & Stone, 1979) was developed on the basis of postulates that differ from those of G theory, and seemed in consequence to offer some alternative, more powerful application possibilities (see, e.g., Bock, 1997;

Embretson & Reise, 2000; Hambleton, 1991; Hambleton, Swaminathan, & Rogers, 1991; van der Linden & Hambleton, 1997). Much has been written about the relative merits of the two different approaches. Each claims appropriateness for specific types of practical application. But each also has underlying assumptions that must be satisfied if the application results are to be valid.

When its relatively strong underlying assumptions are met (unidimensionality, local independence, sample independence), IRT is particularly useful for discriminating effectively at specified levels of difficulty, achieving different forms of equating, detecting item "bias," and managing computer-based adaptive testing sessions. But the assumptions are strong, and while techniques have been devised to check for model fit, their robustness reduces with sample size. The particular advantage of G theory is that it is a sampling model, not a scaling model. Its main interest is its capacity for quantifying and managing sources of error in multifacet situations, under the assumption of an additive linear model of independent effects. G theory is in particular the more appropriate methodology to use when interaction effects are expected. The contributions of main effects and interaction effects to measurement error can be quantified, and an associated SEM computed. IRT, in contrast, treats error as undifferentiated and has no way of dealing with it explicitly, even if local dependence is formally established (Thissen & Wainer, 2001; Yen, 1984). For IRT, interactions between student subgroups and items are evidence of differential item functioning, which is typically addressed by deleting the most offending items during pretesting. The question is: after eliminating potential test items that deviate to some extent from any general pattern of subgroup difference (gender, socioeconomic origin, etc.), to what degree does the set of items that remains validly reflect the ability or skill domain the assessment is in principle focusing on? Interaction effects are important indicators of what it is that a test measures, as noted by Muthén, Kao, and Burstein (1991), in the context of *instructionally sensitive psychometrics*. It seems to us more acceptable in validity terms to retain rather than reject differentially functioning items (unless clearly visibly flawed in some fundamental way), and to reduce their impact on measurement precision by using greater numbers of items in the assessment. In this way the instructional sensitivity of tests can be maximized.

But should we put the two theories in apparent opposition, simply because they appear to be so different, to the point of rejecting one in favor of the other? Could they perhaps be used together, sequentially, to exploit their relative strengths? This is a possibility that several researchers have begun to explore. Verhelst and Verstralen (2001), for example, support the idea of using G theory and IRT sequentially. They write: "The work reported in this chapter is an attempt to find a satisfactory answer to a long-standing practical problem: is there an easy way to handle multifaceted measurement designs in an IRT framework which yields results comparable to those of generalizability theory?" (Verhelst & Verstralen, 2001, p. 104). They suggest a possible path towards this aim: "The challenge is that G theory is concerned with partitioning variance; Rasch is concerned with constructing measures. But once measures have been constructed, variance can be partitioned" (Verhelst & Verstralen, 2001, p. 104).

Going beyond sequential use, model integration is increasingly being recognized as an achievable aim. The multilevel model (Goldstein, 1995; Snijders & Bosker, 1999) has been proposed as a first junction point where IRT and G theory might meet. Indeed, "an IRT model is a mixed-effects model that can be cast in a hierarchical linear modeling (HLM) framework, allowing for the inclusion of covariates and design facets" (André Rupp, personal communication, March 6, 2008). Similarly, the computational procedure typically used by multilevel modeling software can be used to treat ANOVA designs. This is because the equations of the multilevel model are to a large extent additive, the observed result being considered as the sum of a series of postulated effects. The work of Jeon, Lee, Hwang and Kang (2007) took advantage of this parallelism to compare the components of variance obtained by the two theoretical approaches—G theory and multilevel modeling—with balanced and unbalanced data. With balanced designs, the components of variance for each source were the same, and hence the reliabilities of the class means they were focusing on were estimated similarly. This example shows that G theory and the multilevel model can produce concordant results in some cases. It suggests that more encompassing statistical models could be defined, among which both the IRT and G theory models could take their proper place.

This is exactly what Briggs and Wilson (2007, p. 138) meant when, quoting Goldstein (1995), they noted that "both the GT model used for a G Study and IRT models can be conceptualized as instances of multilevel statistical models." They claim that their approach "incorporates the sampling model of generalizability theory (GT) into the scaling model of item response theory (IRT) by making distributional assumptions about the relevant measurement facets" (Briggs & Wilson, 2007, abstract, p. 131). In other words, they derive estimates of the variance components of G theory from estimates of the item and subject parameters obtained within the IRT framework. Their procedure is iterative and uses a Rasch model. It provides simultaneous access to the two kinds of tool, to improve both the sampling and the scaling of items. G theory guides quantitative adjustments (in the number of levels to sample), while IRT provides qualitative information guiding the choice of items for each particular purpose. The authors describe their effort as "a further step toward building a crosswalk between IRT and GT" (Briggs & Wilson, 2007, p. 150), their results proving that they arrived at "the same estimates of variance components and generalizability as would be reached under the more traditional GT approach" (Briggs & Wilson, 2007, p. 151).

The few other empirical studies that have compared the results of parallel IRT/G theory analyses of the same data set show that they arrive at similarly coherent conclusions, albeit for the simplest of measurement models. Bertrand (2003), for instance, using the outcomes of administration of a nine-item attitude questionnaire to 3600 subjects in the PISA 2000 survey (OECD, 2001), carried out a generalizability analysis and two different types of IRT analysis, and discovered through a G-facets analysis that the item that most correlated with the total score was also the most differentiating item in the IRT model, and vice versa. More importantly, he also found corresponding difficulty parameters with the two approaches. Bain, Weiss, and Agudelo (2008) identified the level of ability that a particular item was most informative for, by interpreting its information curve. Thus IRT adds qualitative information to G theory's rather abstract partitioning and quantification of variance.

Other researchers (including Bock, Gibbons, & Muraki, 1988; Gibbons & Hedeker, 1992) have developed multidimensional item response theories, in order to deal with multiple sources of variation, as G theory does (for an application, see, for instance, te Marvelde, Glas, Van Landeghem, & Van Damme, 2006). Marcoulides (1999), on the other hand, presents a latent trait extension to G theory, based on a kind of factorization of the observations, which enables detection of unusual performances, interactions, and inconsistencies.

At this point in time, no final conclusion can be offered as to the outcome of current attempts at unification. One thing is certain: more integrative models will certainly be proposed. Briggs and Wilson (2007) already offer challenging suggestions for development of their method. Hopefully, their endeavor will succeed in uniting a domain that is too fragmented at present. What we can be sure of is that whatever might be the measurement theories of the future, the principles of G theory will remain useful for exploring and controlling sources of variance in diverse fields of application.

appendix A

Introduction to the analysis of variance

Analysis of variance

The ANOVA is a classic statistical technique. It is a procedure with which the total variation in a set of data is decomposed into a number of linearly additive components. When we speak of the ANOVA in the context of G theory, it is in this sense—decomposition, or partitioning, of total variance—that the term should be understood. In this short overview we will illustrate the process of variance decomposition.

Composition of the total variance

Imagine a situation in which $n = 5$ students (S1–S5) are rated in some sense by $k = 4$ different raters (R1–R4), using the same rating scale (ratings running from 0 to 18, for instance). The resulting rating data ($n \times k = 5 \times 4 = 20$ ratings) can be displayed in a two-dimensional grid (Table A1), with rows and columns representing, respectively, the factors Students and Raters. One can calculate the total variance for the 20 scores in this data set by applying the usual descriptive formula, where x_{sr} is the score attributed to student s by rater r, and M is the overall mean score (the average of the 20 individual rating scores):

$$\text{Var} = \frac{\Sigma_{sr}(x_{sr} - M)^2}{n \times k}.$$

The numerator in this expression is the sum of squared deviation scores, or "sum of squares," which is itself a measure of the variability in

Table A1 Between-Student Effect

	R1	R2	R3	R4	Means	Total variance among the 20 rating scores shown here is
S1	6	6	6	6	6	
S2	8	8	8	8	8	$SS(T) = 160$
S3	10	10	10	10	10	$Var(T) = 160/20 = 8$
S4	12	12	12	12	12	
S5	14	14	14	14	14	$SS(T)$: total sum of squares
						$Var(T)$: total variance
Means	10	10	10	10	10	

the data set. We can designate this by $SS(T)$, for total sum of squares. The denominator is simply the total number of data points in the data set. Recall in this regard that there are two expressions for calculating the variance of a distribution, distinguished by whether the denominator is n or $n - 1$ (here, 20 and 19, respectively). The first, the "descriptive approach," is valid in the case where the data set includes all possible elements, or all the elements of interest. The second, the "inferential approach," is used when the aim is to produce an unbiased estimate of the variance of an infinite population on the basis of the data available for a sample. We adopt the first of these approaches principally because it preserves strict additivity for the variances that will be calculated. Now let us consider three different rating outcome scenarios, to see what comprises this total variance.

First case

As a first example, suppose that the four raters rated each of the five students in exactly the same way, that is, each rater gave any particular student the same rating as all other raters. The results might then look like those in Table A1, in which the students are ordered in terms of the ratings given to them, from the lowest rated to the highest rated (mean ratings shown in the final column). The overall mean of the 20 separate ratings is 10. The total sum of squares, $SS(T)$, is the sum of the squared differences between each rating and this mean, that is, $(6 - 10)^2 + (8 - 10)^2 + (10 - 10)^2 + (12 - 10)^2 + (14 - 10)^2 + \cdots$, which is equal to 160. The total variance in the set of ratings, $Var(T)$, is then 160/20, or 8. The Students sum of squares, $SS(S)$, is the sum of the squared differences between each student's mean rating and the overall mean rating, and is thus 40. The Raters sum of squares, $SS(R)$, is similarly the sum of squared differences between each rater's mean rating and the overall mean rating, and is here equal to zero.

It is clear that the variability in the 20 ratings depends entirely on the rating differences that exist between the students (the last column in Table A1). Consequently, this variability can be reconstructed from the variance of the marginal means, using one or other of the following procedures (with k the number of raters):

$$SS(T) = k \times SS(S) = 4 \times 40 = 160,$$

$$Var(T) = Var(S) = 8.$$

Second case

In contrast with the previous example, suppose now that the four raters differ from one another in their degrees of severity. These differences are evident in the last row of Table A2, in which raters are ordered from the most severe to the least severe. Suppose further that the degree of severity of individual raters applies equally to all students. This is reflected, for example, in the fact that the deviation between the mean degree of severity for rater R1 and the mean degree of severity of all the raters $(8 - 10 = -2)$ is identical in the ratings attributed by this rater to each of the students: 4 rather than 6 (mean rating) for student S1, 6 rather than 8 for student S2, and so on. The rater's degree of severity (as, moreover, the degree of severity of all the other raters) is thus the same for all the students (it does not vary from one student to another). The total sum of squares, $SS(T)$, is in this case 210, the Students sum of squares, $SS(S)$, is 40, as before, and the rater sum of squares, $SS(R)$, is 10.

In this case, the variability in the 20 ratings arises from two types of factor, variability among students on the one hand and variability among raters on the other. Moreover, the effects of these two factors are additive, which means that the total variance can be calculated simply by summing

Table A2 Between-Student and Between-Rater Effects

	R1	R2	R3	R4	Means	Total variance among the 20 rating scores shown here is
S1	4	5	7	8	6	
S2	6	7	9	10	8	$SS(T) = 210$
S3	8	9	11	12	10	$Var(T) = 210/20 = 10.5$
S4	10	11	13	14	12	
S5	12	13	15	16	14	$SS(T)$: total sum of squares
						$Var(T)$: total variance
Means	8	9	11	12	10	

the variance among students' mean ratings and the variance among raters' mean ratings:

$$SS(T) = k \times SS(S) + n \times SS(R) = (4 \times 40) + (5 \times 10) = 210,$$

$$Var(T) = Var(S) + Var(R) = 8 + 2.5 = 10.5.$$

The total variance is the combination of two independent components, and so we say that the variability in the set of 20 ratings is due to the differences that exist between the students on the one hand and between the raters on the other. We also say that this variability is "explained" by the effects of these two variance sources: students and raters.

Third case

While the two previous cases are of theoretical rather than genuinely practical interest, this final situation is entirely representative of those that we come across when engaging in research and evaluation. This is because raters typically not only differ in their overall (or average) degrees of severity, but also apply their standards unevenly from one student to another. Thus, a particular rater can be relatively more severe when rating one student than when rating another. We see in Table A3, for example, that rater R1, whose overall degree of severity results in ratings two points lower than the average severity of all four raters (8–10), is even more severe when rating students S2, S3, and S5 (–3) and relatively less severe when rating student S4 (+1).

We infer from this that the total variability can no longer be explained solely by the variability among students and the variability among raters, because these two sources take no account of the phenomenon just described. An additional source of variation needs to be introduced that specifically reflects the fact that raters deviate from their usual overall degree of severity as they rate different individual students. This is the

Table A3 Marginal and Interaction Effects

	R1	R2	R3	R4	Means	Total variance among the 20 rating scores shown here is
S1	4	4	8	8	6	
S2	5	8	9	10	8	$SS(T) = 270$
S3	7	11	11	11	10	$Var(T) = 270/20 = 13.5$
S4	13	6	14	15	12	
S5	11	16	13	16	14	$SS(T)$: total sum of squares $Var(T)$: total variance
Means	8	9	11	12	10	

so-called interaction effect between raters and students (symbolized as $S \times R$, or SR). It is calculated on the basis of the deviations that exist between the scores attributed by each rater to each student and the scores that would be expected if there were no interaction between the two factors (these expected scores appear in Table A2). Thus, working down rater columns in the two tables, we have $SS(SR) = (4 - 4)^2 + (6 - 5)^2 + (8 - 7)^2 + (10 - 13)^2 + \cdots + (16 - 16)^2$, or 60. $SS(S)$ remains at 40 and $SS(R)$ remains at 10, as in the previous example.

To find the total score variance, the variance attributable to the interaction between raters and students, $Var(SR)$, must be added to the student variance and the rater variance, giving

$$SS(T) = k \times SS(S) + n \times SS(R) + SS(SR) = (4 \times 40) + (5 \times 10) + 60 = 270,$$

$$Var(T) = Var(S) + Var(R) + Var(SR) = 8 + 2.5 + 3.0 = 13.5.$$

In conclusion, we can say in a research/evaluation context in which there are two crossed factors at play, the total score variance is the sum of three independent components (three sources of variation). Should the crossed factors be Students and Raters, then the three variance components relate, respectively, to the variability among students (the Student effect), the variability among raters (the Rater effect), and the variability attributable to interaction between raters and students (the Student–Rater interaction effect).

Variance component estimation

A score matrix similar to that in Table A3 (even if typically of larger scale) is the starting point for the variance partitioning process. The scores within the table exhibit a certain total variability that we can quantify with the usual statistical techniques. The aim of the ANOVA is to partition this total variance, attributing to each contributing source that part of the total variance that it "explains," or accounts for. The procedure for estimating the different parts of the total variance has a number of steps. In the first step, sums of squares (SS) are calculated, as before, as are the degrees of freedom (df) and the mean squares (MS) associated with each source of variance. We assume that these three operations are familiar to readers, since they appear within several different statistical techniques, including regular ANOVA, with one or more classification factors, and regression analysis.

The formulas used in these calculations are given below, where x_{sr} is the score given to student s by rater r, $m_{s.}$ is the mean score for student s across the four raters, $m_{.r}$ is the mean rating score for rater r across the five students, M is the overall mean score, n is again the number of students, and k is the number of raters:

$$SS(S) = k \times \sum{}_{s}(m_{s.} - M)^2$$

$$SS(R) = n \times \sum{}_{r}(m_{.r} - M)^2$$

$$SS(SR) = \sum{}_{sr}(x_{sr} - m_{s.} - m_{.r} + M)^2$$

$$df(S) = n - 1, \quad df(R) = k - 1$$

$$df(SR) = df(S) \times df(R)$$

$$MS(S) = \frac{SS(S)}{df(S)}$$

$$MS(R) = \frac{SS(R)}{df(R)}$$

$$MS(SR) = \frac{SS(SR)}{df(SR)}.$$

Starting from the mean squares, the "components of variance" for the sources concerned (here S, R, and SR) are estimated. These components in practice describe the variance of that part of each observed score (each x_{sr}) that is specifically attributable to one particular source of variance. Indeed, it is possible to estimate the sampling error variance caused by the interaction of the facet under study with the other facets. (In general, for designs without repeated measures that have only one observation per cell, this sampling effect is estimated by the interaction of highest order.) After subtracting this variance from the corresponding mean square, we obtain a "purified" measure of the "true" variance resulting from this effect. We call it its "component of variance," the circumflex indicating that it is an estimate of an unknown value.

For the example presented here, the expected effect of sampling fluctuations is estimated by the interaction mean square, $MS(SR)$:

$$\hat{\sigma}^2(SR) = MS(SR) = 5.00,$$

$$\hat{\sigma}^2(S) = \frac{MS(S) - MS(SR)}{k} = \frac{40 - 5}{4} = 8.75,$$

$$\hat{\sigma}^2(R) = \frac{MS(R) - MS(SR)}{n} = \frac{16.67 - 5}{5} = 2.33.$$

Table A4 ANOVA Table for the Two-Factor Random Model $S \times R$, with S and R both Random

Source	SS	df	MS	Estimated variance components
S	160.00	4	40.00	8.75
R	50.00	3	16.67	2.33
SR	60.00	12	5.00	5.00
Total	270.00	19		

The procedure just outlined is the one that is applied when dealing with a completely random effects model (all factors are randomly sampled from populations that are assumed to have infinite size). These "random effects" components are shown in Table A4.

On the other hand, if some facets are fixed, or sampled from finite universes, then the "purification" mentioned above should either not be carried out, since there is no sampling, or should be applied only partially (for finite samples). Computations then yield "mixed-model components." Table A5 presents the mixed-model results that would be obtained from the same data set discussed earlier if we were to fix the differentiation factor (S) and to sample the levels of the instrumentation factor (R) from a finite universe of size 20 (penultimate column).

But these mixed-model components are not well adapted either to the task of partitioning the total variance, because they are not additive: the sum of the mixed-model components of variance is not equal to the total variance. A further correction is needed in order to obtain component estimates that are genuinely comparable and that can therefore meaningfully be combined. The mixed-model components must be adjusted by

Table A5 ANOVA Table for the Mixed Model $S \times R$, with S Fixed and R Finite Random[a]

Source	SS	df	MS	Components		
				Random	Mixed	Corrected
S	160.00	4	40.00	8.75	9.00	7.20
R	50.00	3	16.67	2.33	3.33	3.17
SR	60.00	12	5.00	5.00	5.00	3.80
Total	270.00	19				

[a] The final variance component estimates are those produced after adjusting the random effects components shown in Table A4 to address the mixed-model feature (penultimate column), and then applying Whimbey's correction factor (final column).

application of Whimbey's correction (Whimbey et al., 1967). For all effects (main factors and interactions, whether belonging to the differentiation or the instrumentation face and whatever the sample size) the adjustment involves multiplying the mixed-model component estimate by $(N - 1)/N$, N being the size of the factor population. The correction, which must be applied to all sources of variance, has no effect when population sizes tend to infinity, but can make quite a difference when population sizes are small.

To wind up this brief overview, we note that more detailed information about certain estimation procedures is given in Appendix C, which focuses on the calculation and characteristics of the G coefficient for relative measures (Coef_G relative) produced by EduG.

appendix B

Sums of squares for unbalanced nested designs

A design that involves a nested factor is said to be balanced when the number of levels of that nested facet is identical from one level of its nesting facet to another (e.g., the same number of students in every class, the same number of items in every skills domain, etc.). In terms of the analysis of variance, balanced data sets are more straightforward to process than unbalanced ones. Therefore, when real-life data sets are unbalanced, we generally impose balance on them by randomly eliminating "excess" observations, that is, by deleting student or item records. While this procedure is indeed useful, it also clearly results in some information loss. The loss might be considered negligible if the number of randomly rejected levels of the nested factor is minimal—a handful of items in a 40-item test, perhaps, or one or two students in a class of 25, and so on. In some cases, however, the information loss can be more serious, for example, when it involves excluding seven test items from the 15 that constitute a motivation questionnaire, or 10 students in a class of 25. In this type of situation, experience has shown that the results of the subsequent analysis can vary appreciably from one item subset to another, from one student subgroup to another, and so on.

To address the problem, we first compute the sums of squares in such a way that as much of the original information as possible is exploited, using any appropriate software package. This requires weighting the levels of the nesting facet appropriately, taking into account the number of observed levels of the nested facets that are associated with it. Thus, individual classes would be weighted in terms of the number of students in them. Means, including the overall mean, would then also be calculated on the basis of weighted data. Once we have the weighted sums of squares we can introduce them into EduG, and request a G study analysis using these precomputed statistics in place of raw data.

Let us consider a very typical situation, in which n candidates (of generic name c) take an examination, in which k domains (of generic name d) of competence are assessed using p different test items (of generic name $i{:}d$) to represent each domain. The observation design is $C(I{:}D)$, where C, I, and D represent candidates, items, and domains, respectively. Now suppose that the number of items varies from one domain to another, so that data weighting is needed. The weight associated with each level of domains (level d, for example) will be the ratio between $p_{\min}$, the minimum number of items nested in any of the levels of this factor, and p_d, the number of items nested in the particular level concerned, that is, the ratio $p_{\min}/p_d$. Obviously, the domains level that has the least number of nested items levels within it will have a weight of 1 and the other domains a weight smaller than 1.

The weighting factor appears in the formula used to calculate the means of the candidates across the items and domains, as well as in the calculation of the overall mean. The means so "corrected" will be overlined, to distinguish them from the usual means. The "corrected" mean for candidate c is given by

$$\overline{m}_c = \frac{1}{k \times p_{\min}} \times \sum_d \frac{p_{\min}}{p_d} \left[\sum_{i:d} x_{ci:d} \right]$$

$$= \frac{1}{k} \times \sum_d \frac{1}{p_d} \left[\sum_{i:d} x_{ci:d} \right]$$

where $x_{ci:d}$ is the score for candidate c on item i in domain d, $p_{\min}$ and p_d are as defined above, and k is the number of different domains in the examination.

Similarly, with n being the number of candidates examined, the overall mean is given by

$$\overline{M} = \frac{1}{k \times n \times p_{\min}} \times \sum_d \frac{p_{\min}}{p_d} \left[\sum_{i:d} \sum_c x_{ci:d} \right]$$

$$= \frac{1}{k \times n} \times \sum_d \frac{1}{p_d} \left[\sum_{i:d} \sum_c x_{ci:d} \right].$$

A numerical example

Let us now look at a concrete example, in which 10 candidates (C01–C10) attempt an examination comprising seven test items (I1–I7) from two

Table B1 The Score Grid for a Three-Factor Nested Design with Unbalanced Data

	Domain 1 (D1)			Domain 2 (D2)				Means
	I1	I2	I3	I4	I5	I6	I7	
C01	1	1	1	0	1	1	1	0.875
C02	0	0	0	0	1	0	1	0.250
C03	0	1	1	1	1	0	0	0.583
C04	1	0	1	0	0	1	0	0.458
C05	1	1	1	1	1	1	1	1.000
C06	0	0	1	1	0	1	1	0.542
C07	0	0	1	0	1	1	0	0.417
C08	0	0	0	0	0	0	1	0.125
C09	1	1	1	1	0	1	1	0.875
C10	0	1	1	1	1	1	1	0.833
	0.4	0.5	0.8	0.5	0.6	0.7	0.7	
Means		0.567			0.625			
				0.596				

different skill domains (D1 and D2), with three items (I1–I3) representing the first domain and four items (I4–I7) representing the second. The values in italics in Table B1 are the weighted means, produced using the formulas given above. Each candidate mean in Table B1 is simply the arithmetic average of the candidate's two separate domain means (candidate C01's two domain means are, respectively, 1 and 0.75, giving a weighted mean for this candidate of 0.875). The overall mean is also simply the mean of the two separate domain means, this time calculated over all students and items in each domain. The domain means are 0.567 and 0.625 for D1 and D2, respectively, giving an overall mean of 0.596. All other means are calculated in the usual way.

We move on now to present the algorithms for calculating sums of squares for the different factors and factor interactions:

$$SS(D) = n \times p_{\min} \times \sum_d (m_d - \bar{M})^2$$

$$= n \times \sum_d \frac{p_{\min}}{p_d} \left[\sum_{i:d} (m_{i:d} - m_d)^2 \right],$$

$$SS(C) = k \times p_{\min} \times \sum_c (\bar{m}_c - \bar{M})^2,$$

$$SS(CD) = p_{\min} \times \sum_d \sum_c (m_{cd} - \bar{m}_c - m_d + \bar{M})^2,$$

Table B2 Sums of Squares and Degrees of Freedom
for the Numerical Example

Source	SS	df
C	4.626	9
D	0.051	1
CD	1.376	9
I:D	1.073	4
CI:D	7.323	36
Total	14.449	

$$\text{SS}(CI{:}D) = \sum_d \frac{p_{\min}}{p_d} \left[\sum_{i:d} \sum_c (x_{ci:d} - m_{i:d} - m_{cd} + m_d)^2 \right]$$

$$\text{SS}(\text{Tot.}) = \sum_d \frac{p_{\min}}{p_d} \left[\sum_{i:d} \sum_c (x_{ci:d} - \bar{M})^2 \right]$$

The results for the current example are shown in Table B2, from which we can see that the total sum of squares calculated using the last formula above is equal to the sum of the sums of squares associated with each of the five sources of variance. As to degrees of freedom, variance sources $I{:}D$ and $CI{:}D$ need particular attention, given that the number of items differs from one domain to the other. The smallest number, $p_{\min}$, is the appropriate value to use for representing the number of levels of I (within D); here this number is 3. For facet $I{:}D$, for instance, the degrees of freedom will be $k(p_{\min} - 1)$, exactly as if all domains contained $p_{\min}$ items. In the present example, df$(I{:}D)$ will be $2 \times (3 - 1) = 4$, and df$(CI{:}D)$ will be $9 \times [2 \times (3 - 1)] = 36$ (see Table B2).

Using these results, a G study can be carried out with the help of EduG. The observation and estimation designs must be defined first. Clicking then on *Insert data*, followed by *Sums of squares*, opens a window in which the sources of variance and degrees of freedom corresponding to the observation design are listed. The sums of squares shown in Table B2 can then be entered appropriately. Once the data have been saved, the G study can be carried out in the usual way.

It would be wrong to believe that this procedure is equivalent to that where we randomly reject items to impose data balance. The advantage of this weighting approach is that it quantifies sources of variance while using all of the available variance information. And we are no longer faced with the uncomfortable situation where subsequent component estimates and coefficients sometimes vary in important ways depending which subsample of items they are based on. The downside is that variance component estimates based on unbalanced data sets are generally biased.

appendix C

Coef_G as a link between $\hat{\rho}^2$ and $\hat{\omega}^2$

One of the objectives of this book is to present G theory without too much reference to its mathematical underpinning. Nevertheless, it seemed potentially useful to offer interested readers some technical information about the characteristics of the generalizability coefficient (Coef_G) that we use and its relationships with other indexes, in particular $\hat{\rho}^2$ and $\hat{\omega}^2$.

Coef_G is an index that is applicable in measurement situations where any facet can be fixed, finite random or infinite random. Its generic formula is the following:

$$\text{Coef\_G} = \frac{\hat{\sigma}_D^2}{\hat{\sigma}_D^2 + \hat{\sigma}_I^2}$$

where $\hat{\sigma}_D^2$ and $\hat{\sigma}_I^2$ represent, respectively, the estimated differentiation variance and the estimated instrumentation variance appropriate in the situation considered. In this appendix we overview the procedures used to estimate these variances, procedures that depend on the type of facet sampling employed. We then recompose all possible formulas for estimating Coef_G and compare them to those that commonly appear in the mainstream statistical literature, to validate our use of this generic G coefficient.

In the interests of simplicity, we confine the discussion to a measurement design involving just two facets (although recognizing that three or more facets would be necessary to distinguish some formulas that look identical with only two facets). We start with Case A, in which an instrumentation facet I (for Items) is nested within a differentiation facet S (for Subjects), before considering Case B, where the same two facets are crossed.

Case of one nested instrumentation facet

One point must be made clear from the outset. In a nested design, the items presented to each subject are different. If they were not different, then the design would cross the two facets S and I, a situation we examine later. With nested items, the only possible reliability-like formula that we can use is an intraclass correlation that Cronbach et al. (1963) called $\hat{\rho}^2$ (Mx).

This coefficient is now known as the "index of dependability" (ω) that involves the absolute error variance $\hat{\sigma}^2$ (Δ). We will prove that Φ and "absolute" Coef_G are equivalent.

Elements of absolute Coef_G

The formulas for the differentiation and instrumentation variances, taking into account the type of level sampling employed for the two facets, are given below. In all cases presented under Case A (nested unmatched data), S is the differentiation facet with n observed levels (the subjects), $I:S$ is the instrumentation facet with k observed levels within each subject (the items), MS(S) and MS($I:S$) are, respectively, the mean squares for facets S and $I:S$. For all nested designs, the mean square for I cannot be estimated separately, so that only the formula for absolute Coef_G will be given for each situation.

Infinite random differentiation and instrumentation facets

$$\hat{\sigma}_D^2 = \frac{MS(S) - MS(I:S)}{k},$$

$$\hat{\sigma}_I^2 = \frac{1}{k} \times MS(I:S),$$

(C1)

where MS(S) and MS($I:S$) are the mean squares for facets S and $I:S$, respectively, and k is the number of observed levels of the instrumentation facet $I:S$.

Fixed or finite random differentiation facet with infinite random instrumentation facet

$$\hat{\sigma}_D^2 = \frac{N-1}{N} \times \frac{MS(S) - MS(I:S)}{k},$$

$$\hat{\sigma}_I^2 = \frac{1}{k} \times MS(I:S),$$

(C2)

where N is the population size for the fixed facet S and $(N-1)/N$ is Whimbey's correction coefficient (Whimbey et al., 1967) for facet S.

Infinite random differentiation facet with finite random instrumentation facet

$$\hat{\sigma}_D^2 = \frac{MS(S) - MS(I{:}S)}{k} + \frac{MS(I{:}S)}{K},$$

$$\hat{\sigma}_I^2 = \frac{K-1}{K} \times \frac{K-k}{K-1} \times \frac{1}{k} \times MS(I{:}S),$$

(C3)

where K is the population size for instrumentation facet $I{:}S$, $(K-1)/K$ is Whimbey's correction for this facet, $(K-k)/(K-1)$ is the finite population correction factor for the facet, and $MS(I{:}S)/K$ is the mixed-model correction for facet S when facet $I{:}S$ is fixed.

General formula for the calculation of absolute Coef_G

There are six possibilities that can arise depending on how the levels of the two facets are chosen. They are shown in Table C1.

The following formulas apply in all six of these situations for the computation of absolute Coef_G:

$$\hat{\sigma}_D^2 = \frac{N-1}{N} \times \frac{MS(S) - MS(I{:}S)}{k} + \frac{MS(I{:}S)}{K},$$

$$\hat{\sigma}_I^2 = \frac{K-1}{K} \times \frac{K-k}{K-1} \times \frac{1}{k} \times MS(I{:}S).$$

(C4)

Table C1 Possible Combinations of Factor Sampling Strategies in a Two-Factor Design

Differentiation facet A	Instrumentation facet B
Infinite random	Infinite random
Infinite random	Finite random
Finite random	Infinite random
Finite random	Finite random
Fixed	Infinite random
Fixed	Finite random

Relationship between absolute Coef_G and each of Φ and absolute $\hat{\omega}^2$

Absolute Coef_G and Φ

To compare "absolute" Coef_G to other coefficients in current use, we need an example application for which a Φ coefficient would be appropriate: the research should have the same structure as a G study with unmatched data. The design would nest k different items within each of n levels of a factor Subjects, with each item being a question addressed orally to one and only one subject. The classical analysis of variance model with a single classification factor Subjects could be used to compare these Subjects on the basis of a quantification of their answers. The ANOVA results would allow estimation of the Subjects effect. If this factor is declared random, then the appropriate index to measure its effect is an intraclass reliability coefficient, Φ, whose formula is (Winer et al., 1991, pp. 125–126):

$$\text{Intraclass reliability} = \Phi = \frac{MS_{treat} - MS_{error}}{MS_{treat} + (k - 1)MS_{error}}. \qquad (C5)$$

In our case, MS_{treat} is the between-subject mean square and MS_{error} is the within-subject mean square. (For consistency, the n in the original authors' formula has been transposed into k here. So the coefficient of MS_{error} is the number of items minus 1.)

This formula is also presented by the developers of G theory as formula 41 of their original paper (Cronbach et al., 1963). They describe it as the estimate of $\rho^2(Mx)$ for unmatched data. They derive from it their formula 42, which is the formula given below. (We could use the same terminology as Cronbach and associates, but since the alternative symbol Φ is in current usage we also use it here.)

A very simple transformation of the denominator in expression (C5) results in the following expression:

$$\Phi = \frac{\left[MS(S) - MS(I{:}S)\right]}{\left[MS(S) - MS(I{:}S)\right] + k \times MS(I{:}S)}. \qquad (C6)$$

We can now compare absolute Coef_G [derived from (C1) above and written below as formula (C7)] to this expression for Φ:

$$\text{Abs. Coef\_G} = \frac{\left[MS(S) - MS(I{:}S)\right]}{\left[MS(S) - MS(I{:}S)\right] + MS(I{:}S)}. \qquad (C7)$$

Note the presence of k in the error term of Φ, and its absence in the error term of absolute Coef_G. This is explained by the fact that the error term in the formula for absolute Coef_G represents the sampling variance of a mean of k observations. Otherwise, both formulas are identical for the case where the differentiation facet is infinite random.

Absolute Coef_G and absolute $\hat{\omega}^2$

With the design unchanged, consider now the case where the facet Subjects is fixed. To render our comparison between absolute Coef_G and other statistical indices meaningful, we will have to use, instead of Φ, a kind of "absolute $\hat{\omega}^2$." This expression does not appear in the statistical literature at the moment. We use it here to distinguish the $\hat{\omega}^2$ formula that is applied to independent samples (Case A, with a nested instrumentation facet) from that for the "relative $\hat{\omega}^2$" applied to matched samples (crossed facets S and I, in Case B below).

The size of the between-subject effect can be estimated for an absolute $\hat{\omega}^2$, using the following expression (Winer et al., 1991, pp. 125–126), but with the symbols k and n interchanged to correspond to our definition:

$$\text{Abs. } \hat{\omega}^2 = \frac{SS(S) - (n-1) \times MS(I{:}S)}{SS(\text{Total}) + MS(I{:}S)}, \tag{C8}$$

where n is the number of subjects, $SS(S)$ is the Subjects sum of squares, $SS(\text{Total})$ is the total sum of squares, and $MS(I{:}S)$ is the within-subject mean square, chosen by us as the error mean square. This last point is critical, as it is what makes this coefficient "absolute," the between-item variance being confounded with the interaction subject × item in the within-subject variance. Winer et al. (1991, p. 125) speak only of an "error mean square" without articulating its nature. Their formula can be transformed (keeping this specification) into the following:

$$\text{Abs. } \hat{\omega}^2 = \frac{(n-1)/n \times \left[MS(S) - MS(I{:}S)\right]}{(n-1)/n \times \left[MS(S) - MS(I{:}S)\right] + k \times MS(I{:}S)}. \tag{C9}$$

The formula for Abs. Coef_G can be obtained from (C2), replacing n by N since they are equal for a fixed facet:

$$\text{Abs. Coef}\_G = \frac{(N-1)/N \times \left[MS(S) - MS(I{:}S)\right]}{(N-1)/N \times \left[MS(S) - MS(I{:}S)\right] + MS(I{:}S)}. \tag{C10}$$

Once again, with the exception only of the absence of k in the error term for Abs. Coef_G, we observe in (C9) and (C10) identical expressions for this situation where the differentiation facet is fixed.

We see then that if the differentiation facet is infinite random Abs. Coef_G is equivalent to Φ, whereas if this facet is fixed then Abs. Coef_G is equivalent to a stepped up absolute $\hat{\omega}^2$, where the intragroup variance (a confounding of between-item variance with interaction variance) estimates the error variance.

Case of two crossed facets

With crossed differentiation and instrumentation facets, it is possible to separate the between-item variance and the Subjects × Items interaction variance. Coef_G can thus take two forms, one called absolute, corresponding to the expected Φ, the other relative, corresponding to the expected ρ^2. The two cases will be considered separately.

Essentially, the problem is the same as with Case A, where the instrumentation facet was nested in the other: we must prove now that Coef_G is equal to $E\hat{\rho}^2$ when the facet of differentiation is infinite random and that it is equal to $\hat{\omega}^2$ when this facet is fixed. To have the elements of this proof, we must first establish in the next section the value of the two components of Coef_G, the "true" variance and the "error" variance, for each type of sampling.

Elements of Coef_G

Purely random differentiation and instrumentation facets

The original algorithms for the calculation of a reliability coefficient were developed for the case where the levels of all facets (differentiation and instrumentation) were randomly sampled from within infinite universes (infinite random facets). This is the simplest case, where the differentiation and instrumentation variances that feed into the calculation of (relative or absolute) Coef_G are estimated as follows for a two-facet design:

$$\hat{\sigma}_D^2 = \frac{MS(S) - MS(SI)}{k}.$$

For absolute Coef_G:

$$\hat{\sigma}_I^2 = \frac{1}{k} \times \left[MS(SI) + \frac{MS(I) - MS(SI)}{n} \right].$$

For relative Coef_G:

$$\hat{\sigma}_I^2 = \frac{1}{k} \times MS(SI),$$ (C11)

where, as before, S is the differentiation facet with n observed levels (the subjects), I is the instrumentation facet with k observed levels (the items), $MS(S)$ and $MS(I)$ are, respectively, the mean squares for facets S and I, and $MS(SI)$ is the interaction Subjects × Items mean square.

Fixed or finite random differentiation facet with an infinite random instrumentation facet

When the differentiation facet S (Subjects) is fixed or finite random, with the instrumentation facet I (Items) remaining infinite random as before, the expressions for $\hat{\sigma}_D^2$ and $\hat{\sigma}_I^2$ become

$$\hat{\sigma}_D^2 = \frac{N-1}{N} \times \left[\frac{MS(S) - MS(SI)}{k} \right].$$

For absolute Coef_G:

$$\hat{\sigma}_I^2 = \frac{1}{k} \times \left[\frac{N-1}{N} \times MS(SI) + \frac{MS(I) - MS(SI)}{n} + \frac{MS(SI)}{N} \right].$$ (C12)

For relative Coef_G:

$$\hat{\sigma}_I^2 = \frac{1}{k} \times \left[\frac{N-1}{N} \times MS(SI) \right],$$

where N is the population size for facet S, $(N-1)/N$ is Whimbey's correction for the S and SI components of variance, $[MS(I) - MS(SI)]/n$ is the component of variance for I, and $MS(SI)/N$ is the correction of variance component I, required for a mixed-model design where S is fixed.

Infinite random differentiation facet with a finite random instrumentation facet

Should the differentiation facet S be infinite random and the instrumentation facet I finite random, the appropriate expressions for $\hat{\sigma}_D^2$ and $\hat{\sigma}_I^2$ would be as follows:

$$\hat{\sigma}_D^2 = \frac{MS(S) - MS(SI)}{k} + \frac{MS(SI)}{K}.$$

For absolute Coef_G:

$$\hat{\sigma}_I^2 = \frac{K-1}{K} \times \frac{K-k}{K-1} \times \frac{1}{k} \times \left[MS(SI) + \frac{MS(I) - MS(SI)}{n} \right]. \qquad \text{(C13)}$$

For relative Coef_G:

$$\hat{\sigma}_I^2 = \frac{K-1}{K} \times \frac{K-k}{K-1} \times \frac{1}{k} \times MS(SI).$$

where K is the population size for facet I, $(K-1)/K$ is Whimbey's correction for the S and SI components of variance, $(K-k)/(K-1)$ is the finite population correction factor for I, and $MS(SI)/K$ is the correction of variance component S for the mixed-model design when I is fixed.

Note: The formulas obtained when applying the algorithms presented under (C12) and (C13) are equally applicable when both facets are infinite random. In this case, in fact, terms involving N and/or K become redundant, since they tend to 1 or 0 as N and/or K tend to infinity.

General formula for calculating Coef_G

The six possibilities that can arise depending on how the levels of the two facets are chosen have been presented in Table C1 and will not be repeated. The following formulas apply to all these different cases:

$$\hat{\sigma}_D^2 = \frac{N-1}{N} \times \left[\frac{MS(S) - MS(SI)}{k} + \frac{MS(SI)}{K} \right].$$

For absolute Coef_G:

$$\hat{\sigma}_I^2 = \frac{K-1}{K} \times \frac{K-k}{K-1} \times \frac{1}{k}$$

$$\times \left[\frac{N-1}{N} \times MS(SI) + \frac{MS(I) - MS(SI)}{n} + \frac{MS(SI)}{N} \right]. \qquad \text{(C14)}$$

For relative Coef_G:

$$\hat{\sigma}_I^2 = \frac{K-1}{K} \times \frac{K-k}{K-1} \times \frac{1}{k} \times \frac{N-1}{N} \times MS(SI).$$

Relationship between relative Coef_G and each of expected $\hat{\rho}^2$ and expected relative $\hat{\omega}^2$

We indicated in Chapter 2 that Coef_G is equivalent to a stepped up form of the intraclass coefficient $\hat{\rho}^2$, when the differentiation facet is infinite random and of $\hat{\omega}^2$ when the differentiation facet is fixed. This relation will be proven now for the two-facet crossed design (as before, S is the differentiation facet and I the instrumentation facet). Care must be taken, however, to distinguish absolute and relative Coef_G, which have of course different formulas.

Relative Coef_G and expected $\hat{\rho}^2$

Imagine a grid with n rows representing n subjects and k columns representing k items (all subjects attempt all items). If we want to compare the subjects and estimate the between-subject effect, then we carry out a classical analysis of variance for a repeated measures design (as shown in Appendix A). If the facet Subjects is a random factor, then the appropriate reliability coefficient is an intraclass correlation, with the estimate of the between-subject variance in the numerator and the sum of two estimates: of "true" variance and of "error" variance in the denominator.

This intraclass correlation is presented in the Cronbach et al. (1963) paper as Equation (30), and is referred to as the Hoyt formula. The authors state (p. 152) that this formula "can be recommended whenever matched data are employed in the G and D studies" (which is the case when the S and I facets are crossed). Here is the formula for the relative intraclass correlation, as it appears in the original paper:

$$\hat{\rho}^2_{Mi} = [\text{MS(Persons)} - \text{MS(residual)}]/\text{MS(Persons)}.$$

MS(residual) is defined as the sum of the interaction and residual variances. This formula is also known as the α coefficient. It is particularly simple, but its use is limited to the two-facet case.

If now we recompose the formula for Coef_G using the elements presented in Equation C11, we obtain

$$\text{Coef}\_G = \frac{\hat{\sigma}^2_D}{\hat{\sigma}^2_D + \hat{\sigma}^2_I} = \frac{(\text{MS}(S) - \text{MS}(SI))/k}{(\text{MS}(S) - \text{MS}(SI))/k + (1/k) \times \text{MS}(SI)}.$$

Multiplying the numerator and the denominator by k, and eliminating the term MS(SI) from the denominator through cancellation, we have:

$$\text{Coef}\_G = \frac{\hat{\sigma}^2_D}{\hat{\sigma}^2_D + \hat{\sigma}^2_I} = \frac{\text{MS}(S) - \text{MS}(SI)}{\text{MS}(S)}.$$

It is clear that this formula for Coef_G (infinite random facets S and I) exactly replicates the intraclass correlation described by Cronbach et al. (1963) as $\hat{\rho}^2_{Mi}$. Both formulas give the reliability for the mean of the k items.

Relative Coef_G and expected relative $\hat{\omega}^2$

The fact that Coef_G can correspond to expected relative $\hat{\omega}^2$ when the differentiation facet is fixed can be tested through recourse to a published formula for $\hat{\omega}^2$. Winer et al. (1991, p. 125) offer the following expression (with their n and k interchanged here for consistency):

$$\text{Rel. } \hat{\omega}^2 = \frac{(n-1)/(nk)\big[\text{MS(treat.)} - \text{MS(error)}\big]}{(n-1)/(nk)\big[\text{MS(treat.)} - \text{MS(error)}\big] + \text{MS(error)}}.$$

Using our facet identifiers, and specifying a possible source for the error variance (which was not defined) we have

$$\text{Rel. } \hat{\omega}^2 = \frac{(N-1)/(Nk)\big[\text{MS}(S) - \text{MS}(SI)\big]}{(N-1)/(Nk)\big[\text{MS}(S) - \text{MS}(SI)\big] + \text{MS}(SI)},$$

where MS(S) corresponds as before to the mean square for the differentiation facet Subjects, MS(SI) is the error mean square, N is the size of the Subjects universe, and k is the number of sampled items. Remember that we chose MS(SI) to estimate MS(error).

If we recompose the formula for Rel. Coef_G using the elements presented in Equation C12, we have

Rel. Coef _ G =

$$\frac{\big[(N-1)/N \times (\text{MS}(S) - \text{MS}(SI))/k\big]}{\big[(N-1)/N \times (\text{MS}(S) - \text{MS}(SI))/k\big] + \dfrac{1}{k} \times \big[(N-1)/N \times \text{MS}(SI)\big]}.$$

Two differences appear between the formula above for Rel. Coef_G and that for relative $\hat{\omega}^2$. The first is the presence of the multiplier $1/k$ in the error variance for Rel. Coef_G. The reason for this has already been explained: Coef_G gives the reliability of the mean of the k items, whose sampling variance is k times smaller than the sampling variance of the individual items.

More interesting is the fact that the error variance, MS(SI), is subjected to Whimbey's correction $(N-1)/N$ in the formula for Rel. Coef_G. The formula given by Winer et al. (1991) does not include this correction, because the "error" variance was not actually defined by these authors. Should it be a within-student variance, for instance, then their formula would be correct. But since we defined the sampling error as the interaction variance of the two crossed facets Whimbey's correction becomes necessary. Interactions with a fixed facet were explicitly mentioned in Whimbey et al. (1967). We see here that the EduG algorithm avoids a potential error that is probably frequently made.

To conclude, the general formula for Coef_G embraces both the $\hat{\rho}^2$ and $\hat{\omega}^2$ coefficients, or their equivalents, as special cases among a larger number of possibilities of application. It must be admitted that proof of this has been offered only for two-facet designs, for which particular results can be obtained. For instance, the value of Coef_G happens to be the same whether the differentiation facet is fixed, random, or random finite. This fact has already been noticed by Brennan who wrote: "The size of the universe of generalization does not affect expected observed score variance" (Brennan, 1986, p. 86). But this equality would not be true with phi or with Coef_G for complex multifacet designs. In this case, the use of a formula taking into account fixed facets of differentiation is essential.

appendix D

Confidence intervals for a mean or difference in means

The purpose of this appendix is to establish links between classical methods of statistical inference and the G theory approach. To this end, we are interested above all in the case of relative measurement, in particular as regards the calculation of standard errors (of means or of differences between means) and of confidence intervals (for a mean or a difference between means). We nevertheless offer a short section on absolute measurement, which rarely features within the framework of classical methods, but which can be relevant and useful for adequately treating certain types of problem.

The following points are addressed here:

- Variance analysis: the classical approach
- G theory: the two-facet case
- Fixed or finite random differentiation facets
- Three-facet designs
- The case of absolute measurement.

Variance analysis: The classical approach

Let us consider a regular ANOVA, with a single k-level factor, Groups (G), whose levels comprise independent student (S) samples, or the same student sample (repeated measures). The first might concern analysis of data from a comparative study, where the effects of k different teaching methods on independent random samples of students are being evaluated. The second might be analysis of data from a progression study, in which the performance of a group of students is measured at k different points throughout the duration of a program of study. We assume that the within-groups sample size, n, is constant. Two mean squares can then be calculated: the mean square associated with the main factor, Groups (G), and the residual mean square, or error. The residual mean square is then

- MS(S:G), if the student samples are independent (the mean square describing the variability among students within groups).
- MS(SG), if the same student sample features at every level of the group factor (the mean square now describing the interaction between students and the levels of the G factor).

The results allow us to carry out a significance test on the levels of the Groups factor (comparison of the means of the k student samples concerned, or the k instances of the same sample at different moments). A third mean square, MS(S), can be calculated in the case of repeated measures samples, and this reflects the variability among students for all the levels of Groups. This mean square does not feature in hypothesis testing, but it can be important when estimating ρ^2 or ω^2 (see Appendix C).

In a second-stage analysis we typically move on to establish confidence intervals around a mean or around a difference between means. To do this we need the corresponding standard errors (of the mean or of the difference between means), which we can obtain by separately considering each group or pair of groups. It is also possible to calculate an overall error term (global, pooled) that can then be used to establish all the confidence intervals of interest, as follows:

Standard error of the mean: SE(M)

Independent samples   Repeated measures samples

$$SE(M) = \sqrt{\frac{MS(S{:}G)}{n}}$$ $$SE(M) = \sqrt{\frac{MS(SG)}{n}}$$

Standard error of the difference between means: SE(D)

Independent samples   Repeated measures samples

$$SE(D) = \sqrt{2 \times \frac{MS(S{:}G)}{n}}$$ $$SE(D) = \sqrt{2 \times \frac{MS(SG)}{n}}$$

$$SE(D) = \sqrt{2} \times SE(M)$$ $$SE(D) = \sqrt{2} \times SE(M)$$

The standard error thus obtained (of the mean, or of the difference between means) can be considered as a kind of "average" standard error, and its use can be justified by the fact that, having been calculated on the basis of the whole set of available data (covering all the sample characteristics), it represents in principle a more stable (and hence more reliable) estimate of the parameter we are trying to establish. Confidence intervals are then classically defined in the following way:

Confidence interval around a mean:

$$L_{L/U} = m \pm z_\alpha \times SE(M).$$

Confidence interval for a difference between means:

$$L_{L/U} = d \pm z_\alpha \times SE(D),$$

$$L_{L/U} = d \pm z_\alpha \times \sqrt{2} \times SE(M),$$

where $L_{L/U}$ represents the lower and upper limits of the interval, m is the sample mean (mean of one level of the Groups factor), d is the difference between the means of two samples, $SE(M)$ is the standard error of the mean, $SE(D)$ is the standard error of the difference between means, and z_α is the critical value that captures the appropriate percentage of values within a Normal or Student distribution.

It must be remembered, however, that in the case of repeated measures samples, the use of $SE(M)$ and $SE(D)$, calculated as shown earlier, is controversial. Admittedly, the term $MS(SG)$ does not seem to pose major problems as far as the global variance analysis test is concerned. [Users of the method know that when certain assumptions are not met, procedures exist for appropriately correcting the results (degrees of freedom and mean squares) used for the variance analysis test, notably the procedures of Greenhouse and Geisser (1959), and of Huynh and Feldt (1970).] But the resulting standard errors appear to be particularly sensitive to violations in the assumptions underpinning the method, in particular the assumption of sphericity. [In addition to equality of variance for all the levels of the Groups factor, which in this case are all the student samples, this condition also assumes equality among the $k(k-1)/2$ covariances between pairs of samples.]

For this reason many authors, though falling short of forbidding the practice, advise against using the pooled (global) standard error for establishing confidence intervals, claiming rather that the necessary estimation should be based directly on the data associated with the level, or difference between levels, of interest (see, e.g., Howell, 1997, p. 530). This is indeed how specialist statistical packages, including SPSS, function when required to perform this type of operation, and when they carry out *a priori* "simple contrast" and "repeated contrast" comparisons.

We note finally that the issue surrounding the sphericity assumption does not arise when the main factor has just two levels, because in this case there is only one covariance. To produce confidence intervals here one can adopt the practice used for independent samples: multiply the pooled SE(*M*) by $\sqrt{2}$ to obtain SE(*D*).

Generalizability theory: The two-facet case

A G study involving one differentiation facet and one instrumentation facet bears many similarities to the situation described above in relation to the ANOVA. We can in fact show that if the differentiation facet is infinite random, then SE(Rel.), the relative measurement error provided by EduG, is identical to the standard error of the mean, SE(*M*), as calculated in the earlier section, for independent and for repeated measures samples.

In the comparative study we have *M* representing the differentiation facet "Methods" (in place of the generic "Groups" used earlier), *S* representing the instrumentation facet Students, with measurement design *S:M*. In the progression study *M*, still the differentiation facet, could represent "Moments," with *S* still representing Students, and with measurement design *SM*. We then have

Independent samples	Repeated measures samples
$\text{SE(Rel.)} = \sqrt{\dfrac{\text{MS}(S{:}M)}{n}}$	$\text{SE(Rel.)} = \sqrt{\dfrac{\text{MS}(SM)}{n}}$

As in classical ANOVA, these terms represent the standard errors with which the confidence interval around an estimated mean can be determined, with the standard error of the difference between means once again found by multiplying SE(Rel.) by $\sqrt{2}$. However, we have seen that in the context of ANOVA the standard error calculated in this way can prove problematic if the differentiation facet has more than two repeated measures samples as its levels. The same issues naturally arise within the framework of a G study, and we must take them into account when establishing confidence intervals. There are two principal ways of doing this. As far as we are concerned, the most sensible way forward is to separately estimate the standard error for each level (sample) or for each pair of levels. This can be achieved by appropriate use of EduG, taking advantage of the observation design reduction possibility described in Chapter 3, with which any particular levels of the Moments facet could be temporarily excluded from the data set, with the analysis then focusing on the remaining levels.

The computational steps are straightforward. For example, let us suppose that, in the situation described above, the differentiation facet *M* has three levels (*k* = 3) and that we want to establish a confidence interval for the mean of the first level. To do this, we conduct a reduced design G study, using the data associated with just the level of interest (observation design reduction). We then have a new MS(S). The standard error of the mean under study, the mean for level 1, is then simply the square root of MS(S)/n, n being the number of observations in the level sample. If we now want to compute the confidence interval for the difference between the means of levels 1 and 2, we carry out another reduced design G study, this time using data from levels 1 and 2 only, excluding that for level 3. EduG records the value of the standard error for relative measurement, SE(Rel.), at the bottom of the G study table. Multiplying this by $\sqrt{2}$ gives us the standard error of the difference between the means of the first two levels. We also always have the possibility of appealing to other statistical packages for these computations (this latter strategy being particularly useful if further in-depth data analyses are envisaged, such as tests for sphericity).

Fixed or finite random differentiation facets

The previous section has considered the case where the differentiation facet is infinite random. But the differentiation facet can be fixed, or even finite random. In such situations the results obtained often differ from those given earlier, because several of the variance components contributing additively to the total variance will have been adjusted through application of Whimbey's correction factor (Whimbey et al., 1967), $(K-1)/K$, where K is the size of the facet universe. (See Chapter 2, "Standard Error of the Mean," and Chapter 3, "Interpreting EduG Reports.")

Readers should note that we are still discussing the two-facet case here, and that the situation will be different for a three-facet analysis. With only two facets, when the samples comprising the levels of the differentiation facet are independent, Whimbey's correction is not needed and therefore has no implications for the calculation of SE(Rel.) or SE(D) (see the left-hand formula below). On the other hand, if the study involved one repeated measures sample, Whimbey's correction reduces the size of the standard error (see the right-hand formula below), this reduction being greater the smaller the size of the facet universe (the minimum value is obtained when the differentiation facet is fixed with just two levels).

Independent samples: fixed or finite random differentiation facet	Repeated measures samples: fixed or finite random differentiation facet
$$\text{SE(Rel.)} = \sqrt{\dfrac{MS(S{:}M)}{n}}$$	$$\text{SE(Rel.)} = \sqrt{\dfrac{K-1}{K} \times \dfrac{MS(SM)}{n}}$$

Faced with this situation there seem to be just two possible ways forward when we are dealing with repeated measures samples. (It goes without saying that the following comments apply only to the value of the standard error for establishing confidence intervals and not to the calculation of G coefficients.)

The first way to proceed is to use the standard errors provided by EduG. For the same set of data, but depending on the sampling involved, the value of the SE(Rel.) varies as shown by the formulas below. The two extreme situations—infinite universe on the left, or two-level universe on the right—call, respectively, for a Whimbey correction of 1 (i.e., no correction) on the left and of 1/2 on the right, under the square root. For the intermediate case of a finite universe with more than two levels, the coefficient of reduction under the square root will be between 1 and 1/2:

Infinite random differentiation facet	Fixed differentiation facet, with two levels
$SE(Rel.) = \sqrt{\dfrac{MS(SM)}{n}}$	$SE(Rel.) = \sqrt{\dfrac{1}{2} \times \dfrac{MS(SM)}{n}}$
$SE(D) = \sqrt{2} \times \sqrt{\dfrac{MS(SM)}{n}}$	$SE(D) = \sqrt{2} \times \sqrt{\dfrac{1}{2} \times \dfrac{MS(SM)}{n}}$

The second possibility, which we recommend, is to use SE(Rel.) calculated for an infinite random differentiation facet. The principal argument in favor of this approach is the fact that classical statistical methods always proceed as though the differentiation facet is infinite random. This is the case, for example, when we compare boys and girls, or the results of a pre/post test study, even when the factors concerned (Gender or Moments) are clearly fixed. To obtain the same results with EduG it suffices to rerun the analysis after declaring the (fixed or finite random) differentiation facet as infinite random.

If the value of SE(Rel.) for the infinite random case is used whatever the sampling status of the differentiation facet, it is possible that the width of the confidence interval might sometimes be overestimated, and the precision of the measurement in consequence underestimated. This outcome is not entirely satisfactory. It is nevertheless to some extent defensible, erring on the side of severity rather than leniency.

(In the preceding paragraphs, we seem to recommend direct use of the overall error term calculated by EduG and designated as the relative standard error of measurement. This is simply because we cannot constantly repeat our warning. If we compare more than two samples with repeated measurements, we should check the assumption of sphericity.

If it is not satisfied, the error term should be computed according to the procedures that statisticians recommend for interpreting F-tests in ANOVA.)

Three-facet designs

The examples presented so far all feature designs involving just two facets (a differentiation facet and an instrumentation facet), which readily allows the demonstration of a close parallelism between classical estimation methods (particularly as applied to ANOVA) and G theory. In contrast, as soon as the number of facets is higher than two, the distinctive characteristics of G theory show up. To illustrate them, an example with one differentiation facet and two instrumentation facets will be briefly presented, corresponding to the following two situations:

   a. *A nested design with independent samples*: Differentiation facet: Methods, M, with k levels. Instrumentation facets: Students, S, and Items, I, with n_s students in each method, and n_i items used. Observation design $(S{:}M)I$, for students within methods crossed with items. Different samples of students follow the different methods.
   b. *A crossed design with repeated measures*: Differentiation facet: Moments (of assessment), M, with k levels. Instrumentation facets, S and I, as above. But the Observation design is SMI, with Moments crossed with Students and Items. The same students are assessed at the different moments; thus we have a repeated measures sample design.

It will be clear when considering these situations that a comparative analysis between the two methodologies (ANOVA and G theory) is no longer possible, for two reasons:

   1. The relative measurement error calculated in G theory differs from that calculated in the ANOVA in that, in addition to the source $S{:}M$ or SM, it involves also MI and $SI{:}M$ in the case of independent samples, and MI and SMI in the case of repeated measures samples.
   2. In G theory there is an operation, "variance purification," in which the "residual" mean square is subtracted from the mean squares associated with the other variance sources when the variance components are calculated: $[MS(S{:}M)\text{–}MS(SI{:}M)]$ and $[MS(MI) - MS(SI{:}M)]$ for independent samples; $[MS(SM) - MS(SMI)]$ and $[MS(MI) - MS(SMI)]$ for repeated measures samples. ANOVA uses individual mean squares for the F-tests (see the last section of Appendix A for the way the random model components are estimated).

Despite these differences, the principles and procedures described earlier are also used in designs with more than two facets. For example, as just mentioned, in the three-facet designs considered here SE(Rel.) is calculated by combining the variance components for *S:M*, *MI*, and *SI:M*, when the *M* levels (samples) are independent, and *SM*, *MI*, and *SMI*, when the samples are repeated measures. The formulas below have been precisely obtained by adding the variance components attributable to the three sources considered.

The two possible approaches outlined above are equally applicable in situations where there are several differentiation and/or instrumentation facets: we can either use SE(Rel.) as calculated by EduG on the basis of the given estimation design, or, as we have suggested, we can use instead the unbiased value, found by rerunning the design analysis with the differentiation facet(s) declared as infinite random. SE(Rel.) is usually smaller if one accepts the reduction due to Whimbey's corrections. The difference between the two procedures is clearly shown in the formulas given below, for a finite and an infinite random differentiation facet, and this for the two kinds of designs, nested or crossed.

Independent samples

Differentiation facet *M* (with *k* observed levels from a universe of size *K*), and instrumentation facets *S:M* and *I* (infinite random, with n_s and n_i observed levels, respectively):

Fixed or finite random:

$$SE(Rel.) = \sqrt{\frac{MS(S:M) + (K-1)/K \times \left[MS(MI) - MS(SI:M)\right]}{n_s \times n_i}}$$

Infinite random:

$$SE(Rel.) = \sqrt{\frac{MS(S:M) + MS(MI) - MS(SI:M)}{n_s \times n_i}}$$

Repeated measures samples

Differentiation facet *M* (with *k* observed levels from a universe of size *K*), and instrumentation facets *S* and *I* (infinite random, with n_s and n_i observed levels, respectively):

Fixed or finite random:

$$SE(Rel.) = \sqrt{\frac{K-1}{K} \times \frac{MS(SM) + MS(MI) - MS(SMI)}{n_s \times n_i}}.$$

Infinite random:

$$SE(Rel.) = \sqrt{\frac{MS(SM) + MS(MI) - MS(SMI)}{n_s \times n_i}}.$$

We recommend using the second formula in each case for all probability-based applications, which simply means rerunning the calculations with the fixed or finite random differentiation facet declared as infinite random.

As far as SE(D), the standard error of the difference between means, is concerned, this is always based on the following general formula:

$$SE(D) = \sqrt{2} \times SE(Rel.)$$

without forgetting that the problems outlined earlier with respect to simpler examples (constructing a confidence interval around a mean or around a difference between means, when repeated measures samples are involved) also feature here, in the same way and with similar consequences.

The case of absolute measurement

Another important difference between classical inferential methods and G theory is the fact that the latter features two types of measurement: relative and absolute. As far as the establishment of confidence intervals is concerned, the procedures described for relative measurement are also appropriate for absolute measurement. The absolute error variance is obtained by adding to the relative error expression contributions from (a) the main effects of facets situated on the face of generalization and (b) interactions effects among these. For the commented example described in the previous section, the two types of error comprise contributions from the following variance sources:

	Observation design $(S{:}M)I$; independent samples; differentiation facet M	Observation design SMI; repeated measures samples; differentiation facet M
Relative error	$S{:}M + MI + SI{:}M$	$SM + MI + SMI$
Absolute error	$(S{:}M + MI + SI{:}M) + (I)$	$(SM + MI + SMI) + (I + SI)$

The formulas used in the calculations are those that define the generalization variance. The standard error for absolute measurement SE(Abs.), then, is simply the square root of the total generalization variance. Starting from SE(Abs.) we can also define the standard error of the difference, in exactly the same way as we have done earlier for the case of relative measurement. We therefore have

Confidence interval around a mean:

$$L_{L/U} = m \pm z_\alpha \times \text{SE(Abs.)}.$$

Confidence interval around a difference between means:

$$\text{SE}(D) = \sqrt{2} \times \text{SE(Abs.)},$$

$$L_{L/U} = d \pm z_\alpha \times \sqrt{2} \times \text{SE(Abs.)}.$$

Key terms

Chapter 1

[i] *Cronbach's α Coefficient*

Reliability is often understood as the consistency of the information given by a measurement procedure. Internal consistency formulas compare the results of two or more subgroups of items. α is an internal consistency reliability coefficient developed by Lee Cronbach, based on the average intercorrelation of the items comprising the test.

[ii] *Standard error of measurement*

The standard error can be considered as the most "probable" error that we risk making when we produce a measurement or, more exactly, as the standard deviation of the distribution of all the errors that are made when the parameter is estimated on the basis of a sample. This theoretical distribution is usually assumed to be Normal (bell-shaped), from which we can infer, for example, that the actual errors made will be smaller than the standard error in two-thirds of the cases and greater in the remaining one-third.

[iii] *Confidence interval*

In statistics, there are two, often complementary, procedures for estimating a population parameter on the basis of a sample drawn from the population: a point estimate and a confidence interval estimate. The first approach aims to identify the "most likely" value of the parameter being estimated. The second aims rather to determine an interval on the measurement scale that has a given probability of containing the parameter's true value, under

Normal distribution assumptions. The most widely used interval is the 95% confidence interval. To give an example, the following formula gives the lower and upper limits of a confidence interval:

$$L_{L/U} = m \pm z \times SE(M),$$

where m is the sample mean (the population estimate), $SE(M)$ is the standard error of the mean, and z is the critical value for the chosen level of confidence (1.96 for a 95% confidence interval and 2.58 for a 99% confidence interval).

Chapter 2

[i] *Confounded facets*

In ANOVA, effects are said to be confounded if they are indistinguishable. For example, in repeated measures designs, where there is just one observation in each cell, the highest order interaction effect is completely confounded with the residual variance, and also with any unidentified systematic effects. Such confounding demands special care when results are interpreted. (Note that incomplete block designs and partially confounded designs cannot be processed using EduG.)

[ii] *Error as sampling error*

Readers may be surprised by expressions like "with error" or "without error," which obviously do not correspond to the everyday usage of the term error. It refers only to the effect of sampling, to the variability that statistical laws enable us to predict. "Without error" means simply "not sampled."

[iii] *Dependability coefficients*

Brennan and Kane distinguish the values of these coefficients at the level of the population and their estimates on the basis of observed samples. They give formulas yielding unbiased estimates that EduG follows when computing $\Phi(\lambda)$, but with often untrustworthy results due to chance fluctuations. Note also that the value of $\Phi(\lambda)$ tells us nothing about the *validity* of the chosen cut-score in mastery or grading applications.

[iv] *Spearman–Brown prophecy formula*

This formula estimates the reliability of a test composed of k' items on the basis of its reliability with k items. Let us designate by r_k the test reliability with k items. Let us call n the ratio k'/k. Then the new reliability with k' items is $r'_k = (n \cdot r_k)/[1 + (n - 1) \cdot r_k]$.

In particular, if r_k is the Bravais–Pearson correlation between the two halves of a test; the reliability of the entire test is $(2r_k)/(1 + r_k)$.

^{v} Measurement bias

For us, "bias" refers to a systematic (nonrandom) error influencing the dependability of a measure. Biases are often due to a flawed instrument design and/or to unsuitable or uncontrolled conditions of observation (items not appropriate to the objective, ambiguous formulation of the question, insufficient time left for executing the task, non-standardized instructions, etc.). If the source of systematic bias is identifiable, then it is sometimes possible to address it by intervening in ways appropriate to the situation. For instance, differences in rater severity can be corrected by subtracting the mean score given by each of them (although this does not address the problem of interaction effects involving raters). Or different norms can be used to evaluate students belonging to different categories, if scores appear to be influenced by student origin, gender, age, and so on.

Chapter 3

^{i} Nesting facet on the differentiation face.

Let us consider a G study in which students are to be differentiated. Students are nested in classes, and items appear on the face of instrumentation. All facets are purely random.

In such a situation, the variance between classes can be treated in two different ways:

- It can be added to the between-student variance, if students are compared without considering the class to which they belong (in this case, between-class variance contributes to the between-student variance).
- It can be ignored, if students are to be compared only within their classes.

But as classes constitute the nesting facet, the between-class variance can never be a source of error affecting the differentiation of students (the nested facet). The fact that a class might be sampled has no influence on the precision of the measures for students forming this class. Furthermore, as the students are different in each class, no interaction can be computed between students and classes: the corresponding variance is confounded with the between-student variance.

Consequently, if a nested facet is the object of measurement, thus forming part of the differentiation face, the nesting facets must also belong to the differentiation face. The variance of the nesting facets will be either added to the variance of the nested facets (thus contributing to the differentiation variance) or it will be ignored in the rest of the analysis and will not affect the computation of the generalizability parameters.

[ii] *Redefining a differentiation facet*

Mathematically oriented readers are referred to Appendix D for the reason why a data set should be rerun through EduG to obtain unbiased SEM estimates, should the G study feature a differentiation facet that is fixed or finite random. It is necessary in that case to define the facet concerned as infinite random, because we need to estimate the expected effect of random sampling interaction effects between differentiation and instrumentation facets. Putting aside the SEM, the reliability coefficients computed by EduG are correct, and optimization procedures based on them will be valid.

[iii] *Independent variables*

The residual scores that remain in the cells of a two-way table after the effects of the rows and of the columns have been subtracted are interaction terms that are uncorrelated and thus independent by construction. The comparison of two levels of the differentiation facet is also subject to fewer constraints than overall significance tests comparing all levels at the same time.

[iv] *Sample size of differentiation facets*

Changing the sample size of a differentiation facet could indirectly affect variance estimates, since the more numerous the observations, the more stable the estimates. But a G study computes the error variance affecting each object of measurement and is not influenced by the number of these objects of study.

Chapter 4

[i] *Cronbach's α*

It would be interesting for readers to note that in any design, like this one, that features a single differentiation facet and a single generalization facet, both sampled purely at random from infinite universes, the G coefficient for relative measurement is identical in value to Cronbach's α coefficient.

Even though their conceptual and technical foundations differ, these two approaches are exactly equivalent in this particular case.

ii *Overall error term*

Readers are reminded that the procedure just presented, of computing an average global estimate of the error variance affecting the difference between any pair of levels, is not recommended if the error components are derived from the interaction of the differentiation facet with the instrumentation facets, that is, in the case of crossed rather than nested designs. The validity of F tests should then be checked as the ANOVA model assumes that all the pairs of levels compared have equal intercorrelations (hypothesis of sphericity). Special procedures for this check are offered by specialized statistical software packages, to which readers are referred.

iii *Spearman's* ρ

Spearman's ρ is a correlation coefficient that can be used when we compare variables measured on an ordinal scale. A classical example where it could be used is the case where n individuals are ranked by two observers according to their motivation level, the rank of 1 being given to the most motivated individual and the rank of n to the least motivated.

This coefficient varies between +1 (in the case where the ranks given are exactly the same) and −1 (when one ranking is the exact opposite of the other). A coefficient equal to 0 indicates that there is no agreement (no relationship) between these two rankings.

iv *Problems with* $\Phi(\lambda)$

Since the difference between the absolute and the criterion-referenced coefficients lies in the fact that the latter takes into account the difference between the test mean score and the cut-score, it would seem logical to assume that the absolute coefficient represents a lower limit for the criterion-referenced coefficient when the test mean and the cut-score are identical. This is in fact true at the level of the population, where the exact value of the general mean of the distribution of scores is known. When we take the observed general mean as an estimate of the true mean, we bias the computation of $\Phi(\lambda)$. Brennan and Kane (1977a,b) demonstrated that an unbiased estimator of $\Phi(\lambda)$ could be obtained if the variance of the grand mean were to be subtracted from the differentiation variance. EduG applies this formula when estimating $\Phi(\lambda)$, but when the cut-score and the observed grand mean are close to each other, the value obtained for $\Phi(\lambda)$ becomes smaller than Coef_G absolute, and can even become negative.

This is why two values are offered by EduG in that case, the raw $\Phi(\lambda)$ that relates to this problem and the restricted Phi(lambda), which is equal to Coef_G absolute, the lower limit of $\Phi(\lambda)$ at the population level. Users can choose the value that seems most appropriate in their situation. In the present example, the absolute coefficient has a value of 0.71 and the criterion-referenced coefficient a value of 0.76. The problem just mentioned does not emerge, because the grand mean is quite distant from the cut-score.

References

Abdi, H. (1987). *Introduction au traitement statistique des données expérimentales.* Grenoble: Presses Universitaires de Grenoble.

American Educational Research Association (AERA), American Psychological Association (APA), National Council on Measurement in Education (NCME). (1999, 2002). *Standards for educational and psychological testing.* Washington, DC: American Psychological Association.

American Psychological Association (APA). (2001). *Publication manual of the American Psychological Association.* Washington, DC: American Psychological Association.

Anscombe, F. J. (1956). Discussion of Dr. David's and Dr. Johnson's paper. *Journal of the Royal Statistical Society B, 18,* 24–27.

Auewarakul, C., Downing, S. M., Praditsuwan, R., & Jaturatamrong, U. (2005). Item analysis to improve reliability for an internal medicine undergraduate OSCE. *Advances in Health Sciences Education, 10,* 105–113.

Bain, D., & Pini, G. (1996). *Pour évaluer vos évaluations—La généralisabilité: Mode d'emploi.* Geneva: Centre for Psychoeducational Research of the Orientation Cycle.

Bain, D., Weiss, L., & Agudelo, W. (2008, January). *Radiographie d'une épreuve commune de mathématiques au moyen du modèle de la généralisabilité.* Paper presented at the 2008 meeting of the Association pour le Développement des Méthodologies d'Évaluation en Éducation—Europe (ADMEE-Europe), Geneva, 8pp.

Bell, J. F. (1985). Generalizability theory: The software problem. *Journal of Educational Statistics, 10,* 19–29.

Bertrand, R. (2003). Une comparaison empirique de modèles de la théorie classique, de la théorie de la généralisabilité et de la théorie des réponses aux items. *Mesure et évaluation en éducation, 26,* 75–89.

Bloom, B. S. (1968). Learning for mastery. *Evaluation Comment, 1*(2) (whole issue; French translation *Apprendre pour maîtriser,* Payot: Lausanne, 1972).

Bock, R. D. (1997). A brief history of item response theory. *Educational Measurement: Issues and Practice, 16*(4), 21–33.

Bock, R. D., Gibbons, R. D., & Muraki, E. (1988). Full-information item factor analysis. *Applied Psychological Measurement, 12*, 261–280.

Borsboom, D. (2006). The attack of the psychometricians. *Psychometrika, 71*(3), 425–440.

Boulet, J. R., McKinley, D. W., Whelan, G. P., & Hambleton, R. K. (2003). Quality assurance methods for performance-based assessments. *Advances in Health Sciences Education, 8*, 27–47.

Boulet, J. R., Rebbecchi, A., Denton, E. C., McKinley, D. W., & Whelan, G. P. (2004). Assessing the written communication skills of medical school graduates. *Advances in Health Sciences Education, 9*, 47–60.

Brennan, R. L. (1986). *Elements of generalizability theory*. Iowa City, Iowa: American College Testing Program.

Brennan, R. L. (2001). *Generalizability theory*. New York: Springer.

Brennan, R. L., & Kane, M. T. (1977a). An index of dependability for mastery tests. *Journal of Educational Measurement, 14*, 277–289.

Brennan, R. L., & Kane, M. T. (1977b). Signal/noise ratios for domain-referenced tests. *Psychometrika, 42*, 609–625.

Briggs, D. C., & Wilson, M. (2007). Generalizability in item response modeling. *Journal of Educational Measurement, 44*, 131–155.

Burch, V. C., Norman, G. R., Schmidt, H. G., & van der Vleuten, C. P. M. (2008). Are specialist certification examinations a reliable measure of physician competence? *Advances in Health Sciences Education, 13*, 521–533.

Cardinet, J., & Allal, L. (1983). Estimation of generalizability parameters. In L. Fyans (Ed.), *Generalizability theory: Inferences and practical applications* (pp. 17–48). New Directions for Testing and Measurement, No. 18. San Francisco: Jossey-Bass.

Cardinet, J., & Tourneur, Y. (1985). *Assurer la mesure*. Berne: Peter Lang.

Cardinet, J., Tourneur, Y., & Allal, L. (1976). The symmetry of generalizability theory: Applications to educational measurement. *Journal of Educational Measurement, 13*(2), 119–135.

Cardinet, J., Tourneur, Y., & Allal, L. (1981, 1982). Extension of generalizability theory and its applications in educational measurement. *Journal of Educational Measurement, 18*, 183–204, and Errata *19*, 331–332.

Christophersen, K-A., Helseth, S., & Lund, T. (2008). A generalizability study of the Norwegian version of KINDL in a sample of healthy adolescents. *Quality of Life Research, 17*(1), 87–93.

Cohen, L., & Johnson, S. (1982). The generalizability of cross-moderation. *British Educational Research Journal, 8*, 147–158.

Cronbach, L. J. (1951). Coefficient alpha and the internal structure of tests. *Psychometrika, 16*(3), 297–334.

Cronbach, L. J., & Shavelson, R. (2004). My current thoughts on coefficient alpha and successor procedures. *Educational and Psychological Measurement, 64*, 391–418.

Cronbach, L. J., Gleser, G. C., Nanda, H., & Rajaratnam, N. (1972). *The dependability of behavioral measurements: Theory of generalizability for scores and profiles*. New York: Wiley.

Cronbach, L. J., Rajaratnam, N., & Gleser, G. C. (1963). Theory of generalizability: A liberalization of reliability theory. *British Journal of Mathematical and Statistical Psychology, 16*(2), 137–163.

Denner, P. R., Salzman, S. A., & Bangert, A. W. (2001). Linking teacher assessment to student performance: A benchmarking, generalizability and validity study of the use of teacher work samples. *Journal of Personnel Evaluation in Education*, 15(4), 287–307.

Embretson, S. E., & Reise, S. P. (2000). *Item response theory for psychologists*. New York: Psychology Press.

Fisher, R. (1955). Statistical methods and scientific induction. *Journal of the Royal Statistical Society B*, 17, 69–78.

Gibbons, R. D., & Hedeker, D. (1992). Full-information item bifactor analysis. *Psychometrika*, 57, 423–436.

Gigerenzer, G. (1993). The SuperEgo, the Ego and the Id in statistical reasoning. In G. Keren, & C. Lewis (Eds), *A handbook for data analysis in the behavioral sciences: Methodological issues* (pp. 311–339). Hillsdale, NJ: Lawrence Erlbaum.

Gillmore, G. M., Kane, M. T., & Naccarato, R. W. (1978). The generalizability of student ratings of instruction: estimation of the teacher and course components. *Journal of Educational Measurement*, 15, 1–13.

Goldstein, H. (1995). *Multilevel statistical models*. London: Edward Arnold; New York: Halstead Press.

Greenhouse, S. W., & Geisser, S. (1959). On methods in the analysis of profile data. *Psychometrika*, 24, 95–112.

Hambleton, R. K. (1991). *Fundamentals of item response theory*. New York: Sage Publications.

Hambleton, R. K., Swaminathan, H., & Rogers, H. J. (1991). *Fundamentals of item response theory*. Measurement Methods for the Social Sciences Series. Newbury Park, CE: Sage.

Heck, R. H., Johnsrud, K., & Rosser, V. J. (2000). Administrative effectiveness in higher education: Improving assessment procedures. *Research in Higher Education*, 41(6), 663–684.

Heijne-Penninga, M., Kuks, J. B. M., Schönrock-Adema, J., Snijders, T. A. B., & Cohen-Schotanus, J. (2008). Open-book tests to complement assessment-programmes: Analysis of Open and Closed-book tests. *Advances in Health Sciences Education*, 13(3), 263–273.

Hogan, T. P., Benjamin, A., & Brezinski, K. L. (2000). Reliability methods: A note on the frequency of use of various types. *Educational and Psychological Measurement*, 60, 523–531. Reprinted as Chapter 4 in Thompson (2003).

Howell, D. C. (1997). *Statistical methods for psychology* (4th ed.). London: Duxbury Press, International Thomson Publishing Inc. [Méthodes statistiques en sciences humaines. (1998). Paris: De Boeck Université.]

Huynh, H., & Feldt, L. S. (1970). Conditions under which mean square ratios in repeated measurement designs have exact F-distributions. *Journal of the American Statistical Association*, 65, 1582–1589.

Jeon, M.-J., Lee, G., Hwang, J.-W., & Kang, S.-J. (2007, April). *Estimating reliability of school-level scores under multilevel and generalizability theory models*. Paper presented at the 2007 meeting of the NCME, Chicago, 21pp.

Johnson, S. (2008). The versatility of Generalizability Theory as a tool for exploring and controlling measurement error. In M. Behrens (ed.), Special Issue: Méthodologies de la mesure. Hommage à Jean Cardinet. *Mesure et Evaluation en Education*, 31(2), 55–73.

Johnson, S., & Bell, J. (1985). Evaluating and predicting survey efficiency using generalizability theory. *Journal of Educational Measurement, 22,* 107–119.

Kane, M. T., & Brennan, R. L. (1977). The generalizability of class means. *Review of Educational Research, 47,* 267–292.

Kreiter, C. D., Gordon, J. A., Elliott, S. T., & Ferguson, K. J. (1999). A prelude to modeling the expert: A generalizability study of expert ratings of performance on computerized clinical simulations. *Advances in Health Sciences Education, 4,* 261–270.

Kreiter, C. D., Yin, P., Solow, C., & Brennan, R. L. (2004). Investigating the reliability of the medical school admissions interview. *Advances in Health Sciences Education, 9,* 147–159.

Lord, F. M. (1952). *A theory of test scores.* Iowa City: Psychometric Society.

Lord, F. M., & Novick, M. R. (1968). *Statistical theories of mental test scores.* Reading, MA: Addison-Wesley.

Magzoub, M. E. M. A., Schmidt, H. G., Dolmans, D. H. J. M., & Abdelhameed, A. A. (1998). Assessing students in community settings: The role of peer evaluation. *Advances in Health Sciences Education, 3,* 3–13.

Marcoulides, G. A. (1999). Generalizability theory: Picking up where the Rasch IRT model leaves off? In S. E. Embretson, & S. L. Hershberger (Eds), *The new rules of measurement* (pp. 129–151). Mahwah, NJ: Lawrence Erlbaum Associates.

Marcoulides, G. A., & Heck, R. H. (1992). Assessing instructional leadership effectiveness with "G" theory. *International Journal of Educational Management, 6*(3), 4–13.

Muthén, B. O., Kao, C.-F., & Burstein, L. (1991). Instructionally sensitive psychometrics: Application of a new IRT-based detection technique to mathematics achievement test items. *Journal of Educational Measurement, 28*(1), 1–22.

OECD. (2001). *Knowledge and skills for life. First results from Pisa 2000.* Paris: OECD.

Rasch, G. (1960). *Probabilistic models for some intelligence and attainment tests.* Copenhagen: Danish Institute for Educational Research.

Searle, S. R., Casella. G., & McCulloch, C. E. (2006). *Variance components* (2nd ed.). Hoboken, NJ: Wiley.

Shavelson, R. (2004). Editor's Preface to Lee J. Cronbach's "My current thoughts on coefficient Alpha and successor procedures." *Educational and Psychological Measurement, 64,* 389–390.

Shavelson, R., & Webb, N. (1991). *Generalizability theory: A primer.* Newbury Park, CA: Sage.

Shiell, A., & Hawe, P. (2006). Test-retest reliability of willingness to pay. *European Journal of Health Economics, 7*(3), 176–181.

Snijders, T. A. B., & Bosker, R. J. (1999). *Multilevel analysis.* London: Sage.

Solomon, D. J., & Ferenchick, G. (2004). Sources of measurement error in an ECG examination: Implications for performance-based assessments. *Advances in Health Sciences Education, 9,* 283–290.

Tatsuoka, M. (1993). Effect size. In G. Keren, & C. Lewis (Eds), *A handbook for data analysis in the behavioral sciences: Methodological issues* (pp. 461–478). Hillsdale, NJ: Lawrence Erlbaum.

te Marvelde, J., Glas, C., Van Landeghem, G., & Van Damme, J. (2006). Application of multidimensional item response theory models to longitudinal data. *Educational and Psychological Measurement, 66,* 5–34.

Thissen, D., & Wainer, H. (Eds). (2001). *Test scoring.* Mahwah, NJ: Lawrence Erlbaum Associates.

Thompson, B. (Ed.). (2003). *Score reliability.* Thousand Oaks: Sage.

Van der Linden, W. J., & Hambleton, R. K. (1997). *Handbook of modern item response theory.* New York: Springer.

Verhelst, N., & Verstralen, H. H. H. F. (2001). An IRT model for multiple raters. In A. Boosma, M. A. J. van Duijn, & T. A. B. Snijders (Eds), *Essays in item response modeling* (pp. 89–108). New York: Springer.

Whimbey, A., Vaughan, G. M., & Tatsuoka, M. M. (1967). Fixed effects vs random effects: estimating variance components from mean squares. *Perceptual and Motor Skills, 25,* 668.

Winer, B. J., Brown, D. R., & Michels, K. M. (1991). *Statistical principles in experimental design.* New York: McGraw-Hill.

Wright, B. D., & Stone, M. H. (1979). *Best test design.* Chicago: MESA Press.

Yen, W. M. (1984). Effect of local item dependence on the fit and equating performance of the three-parameter logistic model. *Applied Psychological Measurement, 8,* 125–145.

Author Index

Subject Index